Audio Wiring Guide

Audio Wiring Guide

How to wire the most popular audio and video connectors

John Hechtman and **Ken Benshish**

Focal Press
Taylor & Francis Group

NEW YORK AND LONDON

First published 2013 by Focal Press

Published 2017 by Routledge
2 Park Square, Milton Park, Abingdon, Oxon OX14 4RN
711 Third Avenue, New York, NY 10017

First issued in hardback 2017

Routledge is an imprint of the Taylor & Francis Group, an informa business

Notice
No responsibility is assumed by the publisher for any injury and/or damage to persons or property as a matter of products liability, negligence or otherwise, or from any use or operation of any methods, products, instructions or ideas contained in the material herein.

Library of Congress Catalog Number: 2008925688

ISBN 13: 978-1-138-40661-2 (hbk)
ISBN 13: 978-0-240-52006-3 (pbk)

Contents

Author Biographies

John Hechtman knew early in life that he wanted to record music – it seemed so magical to freeze time, and be able to replay a cherished event. And to be able to improve the event? Enhance it? Astounding!

With this in mind, John started work in audio studios at the tender age of 16. This unorthodox but highly effective approach gave him decades of hands-on experience, when other young people his age were still in school or just crossing over from school to the real-world workplace.

Learning by doing, and working with masters in the field, John became a proficient audio recordist. He has worked with such diverse artists as Joe Cocker, Lionel Hampton, Jim Carroll, Sy Oliver, The Mighty Sparrow, and many, many more. He says that his 'Andy Warhol' 15 minutes of fame occurred when he worked with The National Lampoon Radio Hour with great comedians like John Belushi, Gilda Radner, Billy Murray, and a slew of other stellar talents.

From working in studios, John branched out into working on studios, building them, wiring them, and managing them. Along the way he picked up a first-rate education in every possible problem an audio studio could have, and in finding ways to solve them.

In addition to his recording and tech talent, John is also a creative writer, songwriter, poet, and amateur inventor. He currently works as a freelance computer and audio consultant, with a strong tendency to prescribe Linux as a cure for the ills of both Microsoft and Apple.

Ken Benshish has a passion for life and celebrates this by sharing what he loves most, music and art, through education. He has played drums in the clubs on Bleeker Street in New York City to percussion with the Chicago Civic Orchestra in Symphony Hall. He has an appreciation for the flowing lines and customization that Harley Davidson offers, and has photographed many of his solo motorcycle adventures throughout the US. Ken is the co-founder and director of the iSchool of Music & Art in Port Washington and Syosset, New York, and is fortunate to be able to enjoy what he loves most in life on a daily basis. Ken lives in lovely Times Square, New York City.

Author Biographies

Basic Information

1

Basic information

Introduction

I wrote this book out of an ongoing sense of frustration with bad workmanship in audio wiring. After 40 years in the recording studio business, the studios I've seen that were correctly wired could be counted on the fingers of one hand.

Often the people who had done the wiring were highly intelligent, motivated individuals. But craftsmanship is not synonymous with either intelligence or motivation. True craftsmanship also requires a thorough understanding of the materials you're working with, an understanding that can be gained *only* through experience. In this book I'll be sharing with you the experience I've gained during decades of audio/video wiring.

The *Audio Wiring Guide* (hereafter AWG) is designed for use by both the amateur and the professional. Whether you're wiring a home studio, a PA (public address system) or a commercial multi-track installation, this book will help you do it better, faster, cheaper, and with fewer mistakes. No matter what the size of your wiring project or installation, the AWG provides you with the essential information you need and the techniques to use it.

One of the biggest differences between the AWG and other books is that the steps you need to do for a particular sequence of work are *illustrated* – with photos that look exactly like the wires in your set-up. The instructions are written so you can understand them the first time you read them, no matter what your experience level.

How we're going to do it

Let's take a trial run now to see how it works.

Wiring nomenclature is often ambiguous and confusing. For example, the word 'wire' could refer to any of these:

- The individual copper strands inside a conductor.
- The strands and their insulating jacket.

- The cluster of conductors and the shield layer in a mic (microphone) or other cable.

All very confusing – and for no good reason! So listen up. In every part of this book, I'll use certain terms in specific ways. Here's an example (see Figure 1.1):

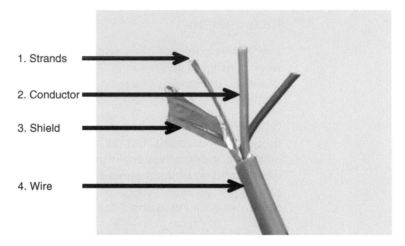

1. Strands

2. Conductor

3. Shield

4. Wire

Figure 1.1   Components of a wire. (1) Copper strands.
(2) Conductor (strands + jacket). (3) Shield – in this case a metalized mylar foil. (4) Wire.

- *Strands* are the individual copper strands of a wire.
- *Conductors* are made up of copper strands that are covered with an insulating jacket (different colors of pliable plastic).
- *Shield* is a metallic, conductive layer wrapped around the inner conductors to reduce noise. It may be a metalized mylar foil, an electrically conductive plastic or actual strands of copper wire that are commonly not insulated.
- *Wires* are made up of the conductors (strands and insulating jackets) in a shield, and commonly surrounded by an outer plastic or rubber jacket.
- A *harness* or *cable* is a collection of wires that are bundled together for a specific purpose.

With me so far? The copper *strands* go into an *insulating jacket* to become *conductors*. Conductors and their *shields* in an *outer jacket* are *wires*. Wires are bundled together to become *harnesses* or *cables* (Figure 1.2).

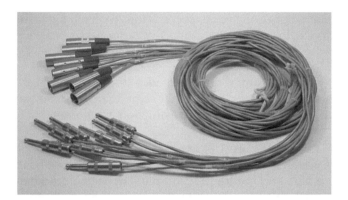

Figure 1.2   Wires in a harness.

The only exception to this rule (there's always something) is the drain (shield) conductor in shielded wire. The drain conductor has no insulation, permitting an electrically conductive contact with the foil or spiral wrap shield along its entire length. So that's why one of the conductors in Figure 1.1 has no insulation on it. In some types of wire, like those with a mylar or conductive plastic shield, the shield and the drain wire are separate. In spiral wrap, and braided shield wire, the drain and shield are sometimes combined in the outer wrapping of copper wire. This is shown in Figures 1.3 and 1.4.

Figure 1.3   Spiral shield wire.

Figure 1.4   Braided shield wire.

Not so difficult, right? All the concepts in the AWG are explained like this – more than once, in fact, so you can follow along easily and understand every point. And the illustrations will show you exactly what I'm talking about.

There are two other common types of wire: these are spiral shield wire and braided shield wire. They do the exact same thing as the mylar foil shielded wire in Figure 1.1, but the shield construction is different.

In spiral shield wire (Figure 1.3), the shield layer is actual strands of copper, wound in a spiral around the inner conductors. The two inner conductors here are the blue and the translucent-over-copper colored items in the picture. The two thinner pale white strands have no electrical function, they are 'packing strands' that help keep the wire round when it's made.

This type of wire is stronger and more noise-resistant then the mylar shield type in Figure 1.1, but it's also larger and costs more. It's flexible and fast to work with, as opposed to the next type of wire I want to discuss.

Braided shield wire (Figure 1.4) offers top notch shielding, and it's very durable. But it's a real pain to work with, because you have to carefully unbraid the shield to connectorize it. Not recommended for the impatient.

Still with me? The three types of wire I've shown you all do the same thing, but they look different, require different techniques, and offer different pros and cons in terms of use. I'm showing all of them to you, because you're likely to encounter all of them in your wiring saga.

A lot of wiring work is like the examples above; the diversity of options available make it seem complicated and confusing. The trick is to see the underlying unity among the options. Three kinds of wire all do the same thing – cool!

If you ever do get confused, just stop, back up a page and read it over – which is a lot easier than hoping for the best, doing it wrong and *doing* it over. Take your time, and the AWG will soon have you soldering like a pro.

Other terms used in this section are explained the *first time* they are used in the text. If you skip a section where a definition is given, or if you forget it, you can look it up in the *online glossary* we've added to the AWG website. We (Focal Press and myself) chose to keep the glossary on the web in order to update it, and to allow more space in the book itself for vital information.

Some sections of the book (like the soldering instructions) are written with deliberate redundancy. If I tell you how to wire a connector, I have to give all the steps in the proper sequence. If you have to flip back and forth in the book to see how a connector is wired, it will only slow you down. So each connector section is designed to be read and followed as a piece of stand-alone text.

A caution, however: the illustrations show the ground wire always connected, since this is how an individual cable would be wired. A star-grounded system would have ground connected at only one end, not both (star grounding is thoroughly covered in a later chapter).

However, be sure you understand the concept of star grounding before doing work on previously installed wiring or starting construction of a new system. The difference in a star-grounded system is that the shield (ground) wire is connected at only one end, rather than both. Connecting shield at both ends of a wire can cause 'ground loops', which induce 'hum' and other types of noise in audio systems. Star-ground installations are always custom-wired and therefore costly – but they radically reduce system noise.

Disclaimer

The techniques explained in the AWG are those that I have used repeatedly in many studio installations with great success. I have made extensive efforts to verify them with other technicians.

But I cannot take responsibility for *your* application of them. Use them at your own risk. I don't mean to scare you (too much), but building and rebuilding studios is part of how I earn a living. I can't offer free tech support, and you may want or need to hire a professional if you can't make these techniques work for you.

However, I'm not asking you to be a guinea pig. Numerous other techni-cians and I have successfully used these techniques over and over. I've built over 20 functional multi-track studios ranging in size from four to 48 tracks. When it was possible to implement the techniques described in this book, they've always worked.

Finally, this book presents high-density instruction clearly and straight-forwardly. You can study it as a textbook or use it as a reference manual during your work. If one picture is worth a thousand words, think of all the text you'd have to read just to get the information in the illustrations!

The tools you'll need

Wiring doesn't require many tools, but you do need some specific items. Let's list them now and talk about them. Necessary tools aren't too expensive, nor are they that hard to find and learn to use. But don't try to start working without them – you wouldn't try to dig a hole without a shovel, would you? Similarly, you need the right tools to wire correctly, rapidly and accurately. If you cut corners here, you'll wind up paying for it many times over – in slow, sloppy work that will need to be redone.

Get a temperature-controlled iron. I've seen more work ruined by people's attempts to use cheap irons than by any other cause. A good temperature-controlled iron will cost only about $60. Getting an iron that is also ESD (electrostatic discharge) safe is a very good investment. That way you're less likely to damage static-sensitive gear when working on it. The blue color of the iron handle in Figure 1.5 is a factory color code by the maker (Weller) that the iron is ESD safe.

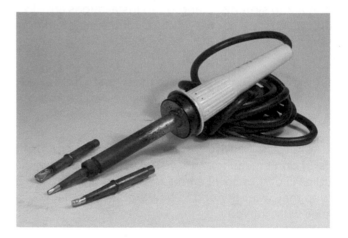

Figure 1.5   Soldering iron and tips.

The best model to get is one with interchangeable tips for a variety of heat ranges and work sizes. That way you can use the iron with a micro-tip for miniature work, and switch to a large tip for heavy soldering jobs. The reason for doing this is that a small tip will be cooled down too much by using it on a large job, even if you have the heat range on the iron cranked all the way up.

Temperature-controlled (T-C) irons work with a magnetic sensor that keeps them at a constant heat. They'll burn out if you place them on a metal holder.

The magnet sees the metal of the holder and keeps the iron on all the time. So if you've made the investment in a T-C iron (and you should), protect it, and yourself, by getting it a proper holder. T-C is the way to go for an iron that stays at the right heat for the job you're doing; I recommend using only T-C irons!

You'll also get better results from your iron by keeping the tip clean with a *soldering iron sponge*. You can use any sponge in a pinch, but iron sponges are heat-resistant, and treated with special chemicals to prevent tip corrosion. You can also buy a tip cleaner that cleans the iron tip without cooling it down like a sponge does.

One possibility is the Radio Shack catalog no. 64-020 'Tip Tinner and Cleaner' (http://www.radioshack.com/product.asp?catalog%5Fname=CT LG&product%5Fid=64-020). Other manufacturers also market a basically identical product (e.g. Multicore TTC1 – http://www.computronics.com.au/ multicore/ttc/). A slightly different approach is the Apogee VTSTC – which also works (http://www.apogeekits.com/solder_tip_cleaner.htm).

Sponge-type cleaners (like the one shown in Figure 1.6) are wetted with water before use. When the soldering iron hits the moist sponge, the tip temperature is lowered – which makes for bad (cold) solder joints. So while sponges are more common than the other types of tip cleaners, they're not really as good – but still totally usable. Just allow a few seconds after you clean the tip (with a sponge) for the iron to reheat.

Figure 1.6   Iron holder and sponge.

Finally, a dry or slightly dampened cloth or paper towel can be used to clean the tip – just be careful that you don't touch the hot iron tip to the towel/cloth for more than a few seconds. This is long enough to clean the iron tip, but not long enough to cause the cloth or paper towel to ignite.

Why all this fuss about keeping your iron's tip clean? A tip that's fouled by excess rosin or old solder won't carry the new molten solder properly to the elements being soldered. In addition, the excess rosin will eventually corrode the tip if left on it for a long period of time.

Buy two pairs of Miller wire strippers (Model 100) or their equivalent. They're available from several sources:

- http://www.kelvin.com/Merchant2/merchant. mv?Screen=PROD&Product_Code=520011&Category_ Code=ELTOWS&Product_Count=5
- http://www.hmcelectronics.com/cgi-bin/scripts/product/5840-0004
- http://www.tecratools.com/pages/servise/wirestrippers.html

Get the kind that has a continuously variable slider to set the strip depth, or a non-preset wheel (Figure 1.7). Some wire strippers have tried to improve this design by providing a wheel with preset strip depth settings. These don't work, because the insulation and wire thickness often falls between the presets.

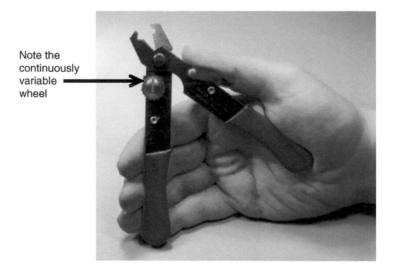

Note the continuously variable wheel

Figure 1.7   Wire strippers.

You don't have to buy genuine Miller strippers, just a kind like them with the slider or non-preset wheel. Imitations are cheap, and available at any good hardware store.

There are dozens of different kinds of wire strippers on the market, with the cost ranging from a few dollars to over $100. The reason I recommend this particular type of stripper is that it's cheap, easy to set up, and works with a broad range of wire and insulation thicknesses.

However, there are a few tricks I'll show you using a small vise and single-edge razor blades that produce faster, cleaner strips than any commercial stripper I've ever seen (more on this later). But even if you use my vise/razor blade tricks, you'll still need a pair of wire strippers for those situations where a vise is inappropriate.

Get three pairs of pliers:

1. One big pair of needle-nose pliers, with a jaw about 2.5 inches long for heavy work.
2. One smaller pair of needle-nose pliers, as pointy as you can find.
3. One pair of ordinary slip-joint pliers, very useful for heavier work.

The two other types shown in Figure 1.8 are optional, but both very useful. The chunky little needle-nose vise grips can act like a miniature vise. The thin hemostatic forceps are great for tiny spaces. Just try to keep them away from all 420 Bohemian loving friends, or you may not get them back.

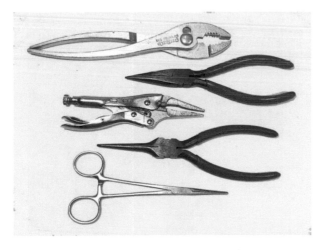

Figure 1.8   Needle-nose and slip-joint pliers.

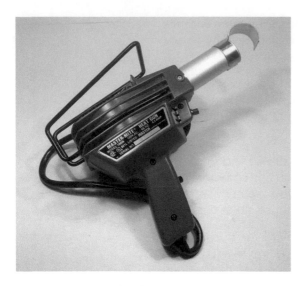

It's often difficult to get really micro-point tools, but you can always grind the tips down to fit your needs.

Heat-shrink is plastic tubing that typically shrinks about 50% in diameter when heated. It comes in a variety of colors, as well as clear. I use a Master-Mite (model 100) heat-shrink gun (Figure 1.9). I find it a good compromise between speed of shrinkage and possible damage to the wire. It's easily available and not too expensive (about $50).

You don't have to use any particular model, but you *must* have a heat-shrink gun to do proper wiring when using heat-shrink. It's not possible to deal properly with heat-shrink without one. A word of **warning** – heat-shrink guns are dandy for lighting cigarettes and starting fires. Make sure your gun is braced so it can't tip over when not in use. And remember to turn it off!

Figure 1.9   Heat-shrink gun.

Tie-wrap guns are used to tighten and cut tie-wraps flush in a single operation. One type looks sort of like a gun, with a long trigger (Figure 1.10). You pull on the trigger, which tightens and cuts the tie-wrap. This gun type is fairly expensive (about $60) but works well.

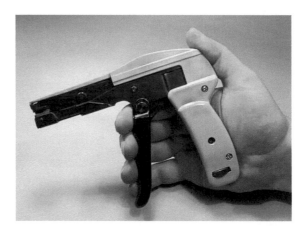

Figure 1.10   Tie-wrap gun.

A cheaper model is a pull-and-turn type, which cuts the tie-wrap when you twist the tool with your hand. These don't work quite as well, but are still acceptable. In an emergency you can pull the tie-wraps tight by hand, and cut them flush with a single-edge razor blade. **Caution:** one way or another you must cut the tie-wraps flush, or the sharp little nubs will draw blood when you handle the cable.

Flush-cutting wire cutters (Figure 1.11) are better than any tie-wrap gun, but they require some practice to use properly. Pull the tie-wrap tight with the jaws of the flush cutters, and then snip off the excess tie-wrap length with them (see the description of wire cutters).

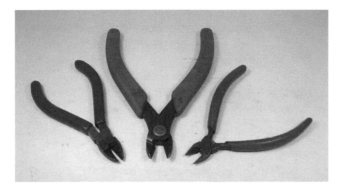

Figure 1.11   Wire cutters.

Wire cutters are also known as 'diagonal side cutters', and are often called 'dykes' (maybe from Diagonal CutterS). You'll need two pairs, large and small. If you use undersized dykes to cut large wire, you can damage or break them.

If possible, try to find flush-cutting dykes for both pairs; they can also be used to flush-cut tie-wraps. Flush cutters do just that – they leave no exposed material at the cut-off point. It's good to reserve one pair of dykes for cutting wire and a second or third (rapidly duller) pair for prying and cutting light metal such as solder on connectors. If you buy a cheap third pair for those dulling, chewing situations, it will extend the life of your better (costlier) dykes.

A small portable vise is used to hold the work in position while you deal with it, and it's a big part of correct wiring. I use something called a 'Vacu-Vise' made by General Tools, but any good small vise will do. If possible, get two vises that are self-supporting – one for you and one for anyone else that does work. If you can't find vises with a base that is already attached, mount them on small pieces of wood so they can be moved around easily.

Figure 1.12   Small portable vise.

The somewhat battered-looking vise in Figure 1.12 has been working for me for 45 years! So you can see that buying good tools pays you back many times over. I thought about painting it or getting a new one, but felt that 45 years of servise gave my old vise the right to wear its battle scars proudly.

Another very useful tool is the 'Third Hand' vise (Figure 1.13). This acts as a companion to your larger vise. The larger vise holds a connector, the Third Hand (TH) holds the wire being soldered to the connector. The TH unit is small – it's got a mini-vise-type weighted base and two movable, adjustable arms, with alligator clips at the ends of the arms.

The alligator 'bites' the wire and holds it in position to be soldered. The model shown is even nicer – it has a built-in magnifier.

Figure 1.13   Third Hand vise.

Professional wire people have trained themselves to work without the Third Hand unit, but I feel it's a real timesaver for both pros and newbies. The TH units are available from many suppliers. One is Shor International Corp. (914-667-1100). The URL for Shor and the TH web page is: http://shorinternational.com/TweezersSlide.htm.

There are as many reasons to buy a volt-ohm-milliammeter (VOM) as there are models to chose from (Figure 1.14). You use VOMs for continuity checks, voltage tests and other types of troubleshooting. Thanks to cheap digital technology, you can get a decent VOM for under $40. Higher priced models add useful features, like an audible beep for continuity testing. The beep feature speeds up your work, so it's worth the difference in cost.

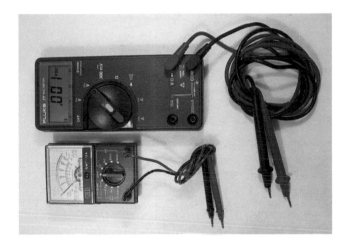

Figure 1.14   Volt-ohm-milliammeters (VOMs).

You can get VOMs that range from shirt-pocket units to laboratory-grade gear that does sophisticated measurements. The price range is equally diverse, from about $15 to over $400. **Tip:** Get a VOM with detachable leads, since the test leads are always the first thing to break. You simply can't check your wiring without a VOM or some other type of continuity tester.

Get one each of flat and Phillips head screwdrivers, in each of the following sizes: large, medium, small and tiny (jeweler's size). For slotted sizes, get a 3.0 mm, a 4.5 mm and a 5.5 mm, plus the jeweler's sizes. For Phillips head, get a no. 1, a no. 2 and a no. 3, plus the jeweler's sizes.

You can get reversible point drivers, so the six types will fit into fewer handles. Using the right size screwdriver avoids damaging the tool and the screw heads in the work itself.

In Figure 1.15 (working down from the top of the photo), you could likely get away with owning the two reversible drivers at the top, along with a jeweler's set like the six small drivers at the bottom. The multi-head driver is handy; the others are shown as examples that might be needed for specific screws.

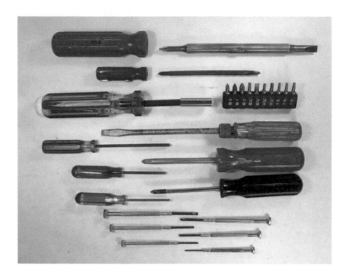

Figure 1.15   Screwdrivers.

Figure 1.16   Solder sucker/solder-wick.

When you need to remove old or excess solder, a solder sucker and/or solder-wick are good ways to do the trick (Figure 1.16). I like the Paladin brand solder suckers; their cold tip model is good and the heated tip model is even better. For many applications I use solder-wick, which is a type of braided wire, chemically treated to absorb solder when heated.

To use it, just place the solder-wick against the excess solder and heat it with an iron.

The solder will flow up into the solder-wick by capillary action. This method of de-soldering is very safe and gentle.

There are a number of special tools you can make with single-edge razor blades (Figure 1.17), so just buy a box or two; later I'll describe the ways you can use them.

You'll use a small metal file to round off sharp edges where you cut away soldered wires. Not only will this avoid puncturing insulation, it will also avoid puncturing your skin when you handle the soldered connectors.

It doesn't really matter what size files you get – three different medium files are shown in Figure 1.18, along with the jeweler's set. Just don't get one that's so large as to be cumbersome when working on small parts. At the top of the picture is a 6-inch machinist's ruler, to give you a better sense of the size of things.

Figure 1.17   Single-edge razor blades.

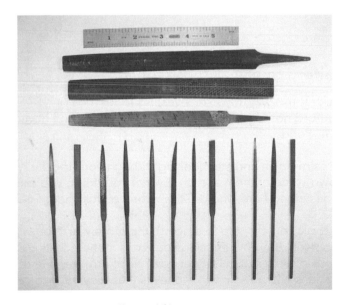

Figure 1.18   Small metal files.

It's pretty obvious – if you're going to measure things, you'll need some measuring tools. Different types of tape measures are shown in Figure 1.19, but a regular ruler is also useful. The best type is an artist's metal ruler about 18 inches long. And don't forget the 6-inch machinist's ruler I showed you along with all those files! If we're going to be measuring things in {1/32} and {1/64} inch sizes, we need a ruler marked off for such tiny amounts. Some kind of precision ruler is essential!

Figure 1.19   Tape measures.

'I can't afford to buy all those tools!'

Well let's put it this way. If you're investing thousands of dollars in your gear, you can't afford *not* to get these basic tools. If you don't have the tools, you can't wire it. Period.

Once you calm down and start looking through your tool kit, you'll realize that you own some of them already. Even if you don't, you can buy *all* the tools I've described here, and use them forever, for under $600.

The more tools you own (up to a point), the more work you will be able to perform yourself, thereby reducing your consultation costs by hundreds or thousands of dollars. Feel better now? Furthermore, one of the reasons that technicians charge so much money is that they have to carry tools, test gear and supplies with them. If a tech knows s/he doesn't have to lug all this stuff to your place, you're more likely to get a break on the price of a service call.

Tools you can build

You can build a number of useful wiring tools yourself. Some of them are cheaper replacements for available devices, and some of them are tools you'll build to do a particular job. Let's look at an example of each.

It's nice to have fancy lights for your work area, but if you're on a tight budget you can create perfectly functional lighting at a fraction of the cost. Got an old Luxo-type light that you can spare for your wiring project? Got a microphone stand? You can combine the two and make a highly functional work light, as shown in Figure 1.20. All we've done here is stick a standard Luxo-type light on a microphone stand. But that cost a lot less than buying a fancy pedestal lamp – and you can easily take it apart when you need the Luxo for your desk and the mic stand for a session.

A home-made jacket slitter is shown in Figure 1.21. Darn, that tiny exposed edge is hard to see! A close-up of the same contraption is shown in Figure 1.22. You need to take care or you'll surely cut yourself with it when you try and use it. OK, got it now? That tiny edge will cut you even better than it cuts insulation on wires.

Figure 1.20 Luxo light on a mic stand.

Note tiny exposed edge at the two corners

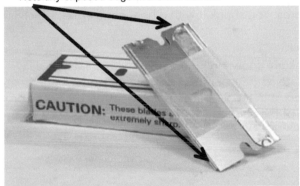

Figure 1.21 Home-made jacket slitter.

Figure 1.22 Close-up of home-made jacket slitter.

Here we've built a jacket slitter by taping two single-edge razor blades together; notice the back (dull side) is in front of the edge (sharp side), except for a tiny exposed sharp edge at two diagonally opposite corners. It takes a bit of experimentation to get just the right amount of exposed edge showing. You want enough edge to score the outer jacket of a wire *without* nicking the insulation of the inner conductors. This is typically {1/64} to {3/64} of an inch of exposed blade. It depends on the insulation thickness, and how hard you push down when you use the jacket slitter.

Once you get the right amount of exposed blade, you can safely score the outer jacket of a wire to strip it off cleanly and easily. And it didn't cost you $50 for a factory-made tool that still has to be set to the correct depth anyway. Practice setting the amount of exposed edge with some scrap wire before cutting the jacket on wires you'll actually be using in your installation.

Later on we'll consider making more tools to speed up your wiring. But right now, let's talk about other materials you'll need to wire.

Other necessary materials

Proper wiring requires certain materials as well as tools. Among them are heat-shrink, labels for your wires, tie-wraps and tie-wrap mounts. Never heard of them? Listen up:

Heat-shrink

This is flexible plastic tubing that comes in various diameters. It's manufactured in such a way that it shrinks 40–50% when heat is applied to it. Heat-shrink is used to insulate wires. You shrink it with a heat-shrink gun, which is a glorified hair dryer.

Different sizes of heat-shrink are shown in Figure 1.23, with a side view in Figure 1.24 to better illustrate its transparency. Here some of the same heat-shrink sizes have been cut to fit in an old pill container. I put a US penny on top of it, to show the scale of the shrink.

Figure 1.23   Sizes of heat-shrink.

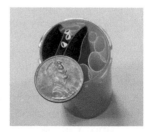

Figure 1.24   Heat-shrink with penny.

As can be seen in Figure 1.25, heat has the effect of shrinking heat-shrink – imagine that! Keep the heat gun moving, don't just let it sit in one position.

This still doesn't quite show you how to use heat-shrink. So let's take a quick look at heat-shrinking the end of a typical wire. This is quite like the 'key frames' in a mini-heat-shrink 'how to' video.

A typical wire end, stripped and ready to shrink, is shown in Figure 1.26. We're going to make it longer than required, so we can later cut it to the exact lengths needed. The

Figure 1.25   The effect of heat.

drain conductor will get a thin piece of shrink, and a larger piece of shrink will go over the breakout of the conductors from the outer plastic jacket.

The same wire with the drain conductor half shrunk is shown in Figure 1.27. Note the difference in diameter between the shrunk and unshrunk portions. It looks good, so we're going to keep shrinking.

Figure 1.26   Wire ready to shrink.

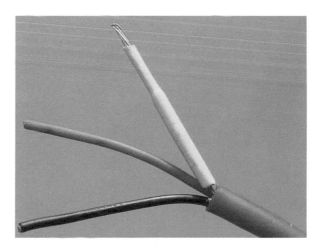

Figure 1.27   Drain conductor half shrunk.

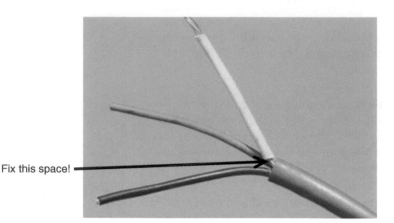

Figure 1.28 Drain conductor fully shrunk.

In Figure 1.28, the heat-shrink is fully shrunk, but note the space that's opened up between the end of the shrink and the outer jacket. To fix it, grip the exposed end of the drain conductor with pliers and push the shrink down toward the breakout from the outer jacket with your fingers. Do this until the shrink is flush with the outer jacket. Don't neglect this step! It's important in making your shrunk wires durable.

Notice how snugly the drain conductor shrink butts up against the outer jacket in Figure 1.29. I was a good wireperson! Now we cap off the wire with a bit of shrink just wide enough to slide easily over the outer jacket. Observe in Figure 1.30 that we leave a bit of the cap-off shrink over the inner conductors and the drain, for strength and insulation.

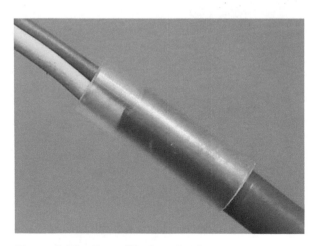

Figure 1.29 Cap-off before shrinking.

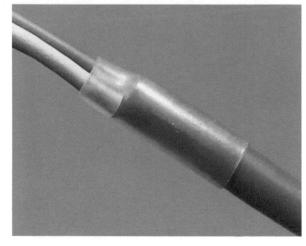

Figure 1.30 Cap-off after shrinking.

A burst of hot air from our trusty heat-shrink gun and the cap-off is complete. The HS gun shown is a medium heat model, but you can still melt insulation, start fires and burn yourself with it.

Some heavy-duty heat-shrink guns are actually paint-stripping heat guns, but **caution:** a paint-stripping gun can *very* rapidly melt wire insulation and start fires. If you have a hard time finding a lighter model heat-shrink gun, a paint stripper is an acceptable alternative. But be careful!

I hope you've now gained a deep spiritual understanding of heat-shrink, and will never again mistake it for a drinking straw or a bit of doll's house-sized garden hose. Used properly, heat-shrink is an invaluable tool.

Wire labels

In the olden days, wires were labeled with numbered markers (Brady markers) that were adhesive backed; then the numbered markers were covered with transparent heat-shrink. All very slow and tedious, and what you wound up with were mysterious numbers that then had to be cross-referenced to the wires and the actual function(s) of the wires in a list or book. And God help you if you ever lost that list/book.

Fortunately, there's now a fast, easy, durable way to identify your wires: self-adhesive wire labels. You can write on them with a razor-point permanent marker, or print on them with an ordinary inkjet or laser printer.

The labels come in various sizes and shapes, but they all have similar characteristics. They're all rectangular, with a small part of the rectangle covered with an opaque white coating. The rest of the label is clear plastic, and the entire label is adhesive backed. To use it, write or print on the white part. There is no standard numbering system for audio wiring, so you'll have to use the markers in a way that makes sense for your installation.

After you peel the label from its backing sheet, stick the white part on the wire *first*, smooth it down, and then wrap the clear 'tail' of the label around the wire and *over* the white part that has the info on it. The clear section acts as 'armor' over the section you've written on. The wire can then be moved or snaked through walls without damaging the new label. Very cool!

The Radio Shack version comes in small packs; the Panduit versions come on sheets suitable for printing. The Radio Shack version is catalog no. 278-1616 (http://www.radioshack.com/product.asp? catalog%5Fname=CTLG&category%5Fname=CTLG%5F011% 5F010%5F008%5F001&product%5Fid=278%2D1616&site=search).

Panduit makes several different label sizes. The 1 inch by 1{5/16} inch size works for almost all the wires I handle. This gives a writeable area of 1 inch by {1/2} inch.

The Panduit number for a slightly larger label (1 inch by 1.5 inch) is S100X150DJ (http://www.panduit.com/products/Products2.asp?partNum=S100X150YDJ&param=361).

A Panduit label that was printed on an inkjet is shown in Figure 1.31. Nice, isn't it? Sharp, clear and fast to work with. This type of label is put on wire as shown in Figure 1.32. Stick the white part of the label on the wire first. Wrap it at a right angle to the wire, so the label's layers overlap equally on the left and right sides of the label. Smooth out any wrinkles as you wrap it.

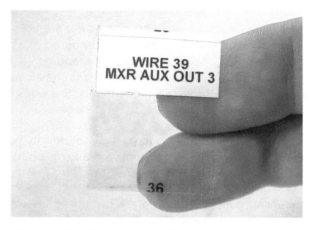

Figure 1.31   Panduit wire label.

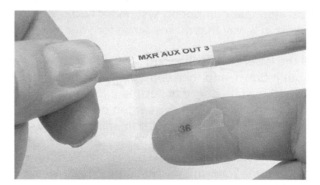

Figure 1.32   Applying label to wire.

Figure 1.33 The finished label.

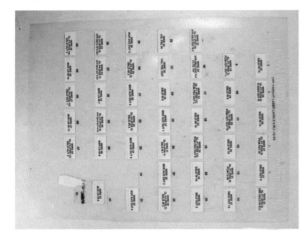

Figure 1.34 Whole sheet of labels.

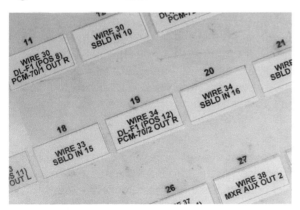

Figure 1.35 Close-up of label sheet.

Figure 1.33 shows the finished label: easy to read, permanent, durable, yet fast to change if the need arises. Print them on an inkjet or laser, or write small runs by hand. An ideal installation tool – use them!

These kinds of labels come on printable sheets. Figure 1.34 shows a view of the labels on their backing sheet. Well, OK, it's not the whole sheet – there are a few missing. But you get my general drift here, right? Still, it's hard to see the labels clearly. Figure 1.35 provides a closer look. Here we can see the labels better. The thin rectangular box around the printing is a line-up guide I made to align the labels on different printers.

Can't stretch your budget for printable wire labels? Or maybe you've run out of labels at 3 a.m. but just *have* to get the job done? No problem – write on white artist's tape or ordinary masking tape and then place it around the wire.

Then cover your writing with some permanent (frosted)-type Scotch tape or clear packing tape, and you have a legible, durable label. Not as good as a store-bought label, but quite workable.

One way or another you *must* label your wires clearly! Don't even think about skipping this step – you'll regret it many times if you do.

Tie-wraps

Tie-wraps are long, straight lengths of plastic designed to curl around wires and lock to themselves, thereby forming the wires into a cylindrical bundle. They come in a broad variety of sizes and lengths, so you can choose the size you need for the wire bundle you're working with.

But bundling the wires is only part of the job; you must also find ways to mount them. For example, let's say you've run and bundled your wires, but they're lying loose on the floor. You constantly trip over them, run your chair into them, and unplug them at critical moments by accidentally pulling on them. And they're beginning to show a bit of wear at the points where you keep hitting them.

How do you get the wires up off the floor and mount them securely? Lucky you – there are lots of ways to do it, from commercial mounts to plain old rope.

Commercial tie-wrap mounts

These come in a variety of sizes and styles. Some are adhesive backed, some have screw-through mounting holes, and some have both. I recommend putting screws through even adhesive-backed tie-wrap mounts, as I've found the adhesive will inevitably fail from the weight of the wires attached to the mount. Here we consider commercial tie-wrap mounts, but don't limit yourself in thinking of ways to support your wires.

In Figure 1.36 we can see the duality of the tie-wrap. They are made flat, but have to curl into hoops when used. Once the tie-wrap's tip (thinner end) is inserted in the clamping (thicker) end, it cannot be released, except for uncommon reusable tie-wraps. However, the details of the tie-wrap mounts aren't very clear, so let's go to our next shots.

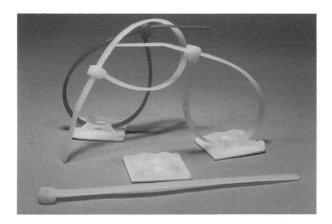

Figure 1.36 Tie-wraps.

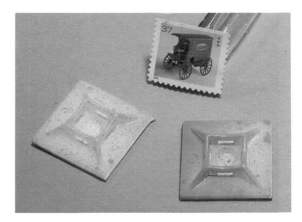

Figure 1.37 Tie-wrap mounts.

The mount details are seen more clearly in Figure 1.37. I had to spray them with blue paint to show up the detail, so don't expect to buy powder blue tie-wrap mounts. These mounts have an adhesive pad, slots for the tie-wraps and a center hole for a mounting screw – use it! I put in a US postage stamp to show how big a typical mount is.

Those painted tie-wrap mounts are seen with a tie-wrap in place in Figure 1.38. If this is still unclear, buy some wraps and mounts, play with them, and you'll understand soon enough.

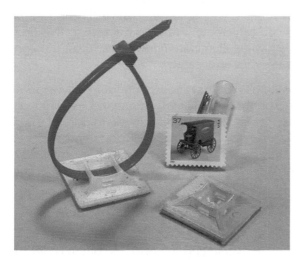

Figure 1.38 Mounts/tie-wrap/stamp.

You can support wire in any manner that doesn't cause sharp edges to contact the wire's insulation. You can use anything that will stay up, do the job, and not abrade the wire being held up. For example:

- Wood cut to fit.
- Metal or plastic plumbing fittings (pipe holders work perfectly).
- Electrical hardware for EMT.
- Other wire.
- Any combination of the above.
- Rope. One of the best low-budget solutions is rope of various thicknesses – non-abrasive, non-conductive, cheap and easily available. Cut short lengths and tie them on at suitable points.

I include no illustrations of the myriad ways to support wire – they'd fill another book. Just let your budget and available materials guide you in supporting your wires. Now let's move on to lighting your work.

Lighting your work

Proper lighting is critical to accurate, high-speed assembly. Most people don't think about this, and just begin work with no attempt to light it properly. When I was a production manager for Eventide, I investigated the effect of lighting on worker efficiency.

A number of well-documented studies show that you can double or triple a worker's productivity with proper lighting. The problem is that to double productivity, you must increase the illumination level four-fold or more. This means that you and your fellow workers will do best in as bright an environment as you can create. Wiring is delicate, precision work, and easily leads to eye strain when illumination levels are low. Poor lighting will also drastically increase your error rate.

This doesn't mean that you have to work under Hollywood Klieg lights. But you do need bright overall illumination, and spotlighting on each work area. I like to use some central lighting in the middle of the room, and scoop lights over each workstation. The scoop lights are cheap and easy to position, since they come with clamps. Even better than scoops are Luxo-type lamps, with a swivel arm and a clamp base (remember how you put one on a microphone stand a few pages ago?).

Organizing your work

It may sound like I'm overstating the obvious, but in order to work, you need a *place to work*. You also need places to store raw materials, documents and tools. If you don't put some effort into organizing the space you work in, it will become a sort of horrid tool salad, and whatever you need most will be lost underneath everything else. You will waste hours of frustrating, counter-productive time looking for misplaced supplies and tools. Now why would you want to do that?

It's not hard to make a good work space. You can lay plywood over saw-horses to make a temporary work table. Use crates, office furniture and whatever is at hand to create workstations. Delicate tables can be covered with heavy cardboard to create a workstation.

How you create your workstations depends somewhat on the size of the job, and where you are doing it. Here are two examples:

While wiring an eight-track studio in a client's home, I put a large folding table in his living room. The wiring was done on the table, and supplies were stored underneath it. By designating a specific work zone, I was able to wire his studio without intruding on the other areas of his apartment. I had to convince him to leave the table up, with tools on it, for the duration of the job. Not having to set up and break down my workstation each time I was there cut dozens of hours off the job.

At the other extreme, when we wired Power Play Studios, I had the carpenters construct work tables out of 4 inch by 8 inch sheets of plywood, mounted at desk height. Each work table provided enough space for four workstations. The legs at the corners were extended above table height, to become columns that I put scoop lights on. Each work table was then self-lit from four directions at once. Power strips were mounted on each table and then recycled into the installation when the work was done. The plywood itself, not being cut, was also recycled at the end of the job.

One way or another, you must create workstations with lighting, air and power for each worker. You also need a place to store materials and documentation.

You *must* also provide adequate *ventilation*. Human beings are mammals, and lack of oxygen will create brain-dead behavior and cause your error rate to soar sky-high! Solder fumes are somewhat carcinogenic, and no one wants the next studio they wire to be the last.

Commercial fume-extraction devices are ridiculously expensive, but a few open windows and some well-placed fans will do almost as well, and not break your budget.

Another important consideration: *containerization*. If you have a paper bag with 50 XLR connectors in it, you will soon have a hole in the bag from the weight of the connectors. You need to put them in a more durable container – a small cardboard box, a plastic salad jar – whatever works and lasts.

The disposable plastic containers with lids used in delicatessens are ideal for containerization of parts. They are cheap, easily available, transparent enough to show what's in them, and come in a variety of sizes. The lids are useful to keep supplies in place, and lidded containers can be stacked.

What you use to hold your supplies depends on two things: what you can easily get, and your own ingenuity. But whatever you do, you *have to do it!* Factories spend good money to organize tools and inventory. They don't do it out of altruism or consideration for their workers. They do it for the sake of productivity, knowing from bitter experience the costly results of inadequate organization. Setting up a wiring job is like setting up a miniature factory – your need for personal organization is as great as that of any large company. Only the scale of operation is different.

Organizing your information

Just as important as organizing your work is organizing your information in a logical and systematic way. Most jobs I've worked on have suffered from lack of organization in documentation. Important schematics were lost, parts lists were mislaid, and a great deal of totally avoidable confusion and hassle was created.

Most people start a job by just piling into it. They don't take the time to think about what they will need or how to do it. Result: a tremendous amount of wasted time and labor. Work has to be redone because it is wrong, and the job waits for materials that are not on site. Worker productivity is low, worker morale isn't much better, and quality control often non-existent.

In this section, I'd like to talk about planning and setting up a job to do it easily and accurately. Start with a meeting, often several, of all the people involved with the project. The purpose of the first meeting(s) is to define *goals*, *methods* and *responsibilities*.

Let's break these terms down. The *goals* of the project are what you hope to accomplish. For example, if you're wiring a 16-track studio and have a low budget, you don't want to plan the wiring for a 24-track upgrade.

You want to be able to plan a job realistically, with the budget you have, for the results you want.

Our hypothetical 16-track studio could go bankrupt trying to finance a big 24-track upgrade too soon. Be sure to work with your existing equipment!

Methods are how you will reach your goals. Are you going to recycle the old wiring or start from scratch? What type of wire should you use and how much should you buy? How are you going to number the wires to keep track of them? Where are the wires going to run – in the floor, in a ceiling, in a trough?

There is a long list of vital questions to answer before you're ready to start wiring. We'll go into these later, in more depth. For now, I only want to use them as examples of methods of work.

It doesn't do any good to have the wires numbered and connectorized correctly if the wires are too short to reach from one piece of equipment to the next. So wire lengths have to be measured (and remeasured) before they are cut. This is what I mean by *methods*.

Responsibilities are just that. *Who* is going to do *what* part of the work and *when* is it going to be done? A large wiring job is far too much work for one person. But if all the workers' efforts are not coordinated, the result is confusion – and slow work that ends up being of poor quality.

Sound way too elementary? You reject these considerations at your peril. Many times I've had to rework jobs because the goals, methods and responsibilities were not clearly defined at the beginning. When jobs go well, be they big or small, they do so because of proper planning. Success is never an accident.

A good example is an installation I did with several other people for Power Play Studios in Queens, New York. Our crew had to design the wiring for a studio that was not yet built. We had to conceptualize the existence of a room, and placement of equipment that was not there!

We spent the first two months of the project making measurements, finding/ ordering materials, designing the wiring scheme, and creating the work area. We took over the upper floor of the building the studio was being constructed in and turned it into a temporary factory. We created work zones, wiring tables, documentation, and feedback questions to the studio owner, all before we ever cut or soldered a wire.

When it came time to do the wiring, we had a clear idea of how to do it, an efficient and well-lit work area, and all the materials we needed. Each worker brought his/her personal tools, and specialized tools were purchased out of the construction budget.

As we progressed through the job, we had space to store finished cables ready for check-out, without interfering with our work zones. We could wire new cables, put them in a pile to be checked, have one worker check and one worker connectorize – all without interfering with each other.

We consulted with the carpenters and acoustic designer, to make sure that there were no built-in obstructions which would interfere with the placement of our wiring in the room. Troughs and tubing (conduit) were added to the construction details, to permit wire runs from one area to another.

It was a race in wiring Power Play to see who would finish first, the carpenters or the wiring crew. The wiring crew won, with two weeks to spare! And when it came time to lay the wiring in and test it, *it all worked*. This is the only time I have ever seen a job check out 100%, and I attribute that success to our careful planning and execution.

Now that we've talked a bit about the general concepts of job planning, let's get down to the specifics you'll need to perform the job.

Job survey and information breakdown

You need a lot of different kinds of information to perform a job correctly. At first it seems almost overwhelming. How much equipment do you have? What type is it? What size is it? How much power does it draw? What are its input and output configurations and impedance? All these questions and more become vital to you. The easiest way to deal with all this is to organize it into lists and drawings.

You *must* have some central place to store the information you gather. I like to use a three-ring binder for documentation. By using a binder, you can insert or delete items, make copies, and organize parts and purchase lists. This will not only help you to do a better job, but will also permit you to keep accurate track of your installation expenses.

For lack of a better name, I call the three-ring binder a 'work log'. You can call it a 'rutabaga' as long as you create and use it. *The work log and documents **you** create will become your single most valuable tool for doing an installation correctly!* I cannot overemphasize this critical point.

How well you gather, and organize, your information is typically the key factor in doing work well, or poorly, rapidly or slowly. This is *so* important that I'm going to go into some details now, and more later, to avoid giving you a mental meltdown from information overload.

Let's talk first about the kinds of information you need to find out. Some of it is best dealt with by listing it, but some of it is easier to gather in a graphic form. For example, suppose that you want to figure out where to put the equipment in a control room.

You've already done your homework, and measured the size of the room, where the windows and doors are, and figured out the traffic flow (where people walk). You've even measured each piece of equipment and made a list of the information.

Now, you want to combine the information you have about the room and the equipment. It would be nice if the way you combined it permitted you to experiment with different placements of the equipment in the room. The question is: how do you do this?

The easiest way I've found is to make a scale floor plan. Take the measurements of a room, and some graph paper (several websites let you download free graph paper in various sizes and print it out).

Establish a relationship between the size of the squares in the grid and the measurements of your room; you want to draw a floor plan of the room, in scale, on the size of paper that you have.

That sounds really complicated, but all I'm asking you to do is to make sure your picture is the right size to fit on the paper. If you draw a room to scale, but one corner runs off the paper, it's hard to figure out what goes in that corner. This is another example of something you might think of as common sense, until you've seen some people's ideas of floor plans.

The concept of 'a scale drawing of your room' is an example of graphic information. It's not a list, even though it's drawn from the information you have compiled in lists. You don't have to be an artist to make a good scale drawing. All you need is:

1. A tape measure.
2. A ruler.
3. Some regular 8.5 inch by 11 inch ruled paper.
4. Some 8.5 inch by 11 inch graph paper.
5. A flat, smooth tabletop.
6. A few pencils with erasers.
7. The will to do it!

I'm sure a lot of you are thinking, 'All this scale drawing stuff sounds like a lot of work. Can't I just put the equipment in place and get down to wiring it?' The answer is: yes and no.

Often, the best arrangement of equipment is not the first one that occurs to you. It's easy to overlook needed room for expansion, work surfaces or traffic flow. Equipment is heavy and delicate; the less you move it, the happier both of you will be.

If you've wired it before you move it, you may face the unhappy prospect of major rewiring when some, or all, of your harnesses don't reach.

Professional studio designers work from floor plans or blueprints. What I'm trying to do is make you work like a professional studio designer in creating your own room. Like most things in life, you get out of it what you put into it – so take your time.

OK, I've convinced you to make a scale drawing of the floor plan you're working with. It's a top view, looking down. You've drawn in any windows, radiators and doors, showing the opening direction of any doors. The floor plan should also indicate available AC outlets and the location of any troughs. What's next?

Now take the measurements you made of the equipment and make scaled shapes of them on a separate piece of graph paper. Cut them out and put them on the floor plan. These scaled shapes can now be rearranged on the floor plan until you find the optimal position for all of them. Be sure to take into consideration things like, 'If I put this item in this spot, will I still be able to open that door completely?'

A scaled floor plan takes time, true, but you can't wire successfully without a clear idea of where things go. And it's just one element in a successful studio information plan. Let's list the other elements now, and describe each one.

Information needed for studio installation

1. *Floor plan.* Top view of the room showing equipment, obstructions, windows, doors, power and troughs.

2. *Equipment lists*. The information needed here can often be combined into one list, with the following categories:
   a. Physical dimensions
   b. Power draw
   c. Input/output impedance
   d. Input/output connector type
   e. Mounting requirements.
3. *Expansion requirements*. Space/power/wiring requirements for new equipment or for gear to be brought into the room on a temporary basis.
4. *Power survey*. Available power in the room. Room power requirements and options in running more power into the room.
5. *Air-conditioning survey*. People and equipment both need air and heating/cooling. What will be needed to provide this for both control room and studio area, as well as other work zones?
6. *Soundproofing survey*. How much soundproofing do you need to get along with your neighbors (or parents) and not get evicted? What treatment do you need to prevent outside noise from intruding into the room?
7. *Acoustical survey*. How well can the room serve for recording and listening back to music? What treatment is needed to optimize the room?
8. *Architectural survey*. Is the building strong enough to permit soundproof construction? Does the building itself have enough power coming into it for your needs? Where are the plumbing lines located to provide water/bathroom/kitchen facilities? Location of cold water pipes or 'I' beams is important in establishing ground. Availability to earth is important for driving ground spikes.

'Aww, gee, do I have to?'

If this sounds like a lot of work, it is. Of course, you can have it done for you by professionals (if you have rich parents or have won the lottery recently). Or you can do it yourself, for a fraction of the cost.

Some of you may be saying, 'This is supposed to be a book about *wiring*! Why are you talking about all these other things?' The answer is, all these things are related. If there is no space to put the wires in the room, you can't wire it. Audio wiring needs to be kept away from AC power runs, and equipment should be connected to separate power legs from those used for air-conditioning or lights. So all these aspects of studio construction have to be considered, however briefly, before you can begin a successful wiring job.

You don't have to find the information in the order of my list. Start with what you know and what you can find out easily. You don't have to be an acoustic designer to do an acoustic survey. You don't have to be an electrician to do a power survey. The same holds true for the other items. You don't have to be an architect or an air-conditioning expert either.

Most of the information required you can research yourself, especially for a small or home installation. Then if you do need to call in professional help, you will be able to talk to them in an informed manner and reduce your consultation costs.

For example, let's create a hypothetical home studio and see how the information I've listed would look. I always like to think of the person I'm designing for, so let's make up a hypothetical person too. We'll call him Iggy; he's a young musician, a college student living at home. Now let's think about Iggy's studio.

Information for Iggy's imaginary installation

1. *Floor plan.* The studio is in Iggy's bedroom. He will have to make allowance for the space taken by the bed and dresser, and allow room for the door to swing into the room.
2. *Equipment list.*
    a. Tascam DA-88 eight-track digital recorder (Iggy got it used – like the one you can see at http://www.bcs.tv/store/prod_detail.cfm?eq_id=423209)
    b. Mackie 1642VLZ3 mixer (you can see this mixer at http://www.boyntonproaudio.com/product-p/1642vlz3.htm)
    c. Alesis Quadraverb – digital delay
    d. Yamaha Rev 500 – digital reverb
    e. Sony PCM 2300 – DAT recorder
    f. Philips CD Recorder CDR-770 – stand-alone CD recorder
    g. Yamaha DX7-II synthesizer
    h. Roland JV-880 – synth module
    i. Hafler DH 220 – power amp
    j. Polk Model RT 25i – speakers.

Those among you who work with DAW equipment (Digital Audio Workstations) will notice immediately that Iggy's studio is all discrete equipment. No DAW – at least not yet.

Iggy has a laptop and does editing with Audacity. Sometimes he manually flies tracks from the laptop back into the eight-track – he's actually gotten pretty good at it.

And he'll mix directly to the laptop, instead of to DAT on occasion, for ease of sequencing.

He's saving up for a good computer-based DAW system that will complement his collection of discrete equipment. He'll also wire a patch bay – that's why he bought a copy of the AWG and met me. For now, he uses his collection of well-chosen, robust gear. He bought most of it used, so he got a lot of bang for his buck.

Remember, this is a hypothetical studio, so I'm not going to list the physical dimensions of the equipment. I would like to talk, however, about the power draw, impedance and connector type of the gear, as well as the mounting requirements.

Power draw for equipment is sometimes given in watts but often given in amps or fractions of an amp. Don't let it worry you; Ohm's law to the rescue! One of the permutations of Ohm's law states that 'amperage × voltage = wattage'. So if a unit draws 0.5 amps and runs on 120 volts, the calculation is 0.5 × 120 = 60 watts. The voltage and current ratings of equipment are usually somewhere on the back of the unit.

The reason for calculating power draw in watts is that it's the easiest way to find how much total power you need, and have available. A 15-amp breaker on a 120-volt line (US voltage) can provide 1800 watts (15 × 120 = 1800).

It's the same formula; we're just using the ratings for the fuse or circuit-breaker to find the total available wattage. All you have to do is find the total draw in watts of your equipment and convert your fuse/circuit-breaker ratings into watts. There are other ways to find total power draw, but this one is easy!

The input/output impedance of equipment is generally listed in the manual. *If you don't have manuals, try to get them.* It will save time for both you and the techs who work with you – hours, and sometimes days, of time. Trust me. Manuals are sometimes downloadable from the manufacturer's website.

Connector types are in plain sight, so count them up and make lists. So many XLR males, so many females, etc. Be as specific as possible as to what type of connectors you have and need. Don't confuse mono with stereo {1/4} inch plugs/jacks – they do different things.

Mounting requirements include not only which units are rack mount, but what gear would be best placed at a special height, or distance, from the operator.

An example would be the DA-88 recorder, which should be at the correct height to see the meters, and for ease of operation when controlling it from the recorder and not the remote. For this studio, I'd like my carpenter friend Joe to build a special wall unit for some of the equipment. I'll put the DA-88 eight-track on a typing table, next to Iggy, and have him face the wall unit which holds all the other gear.

Now back to our information list:

3. *Expansion requirements*. Iggy's friend Fred wants to bring his synth, electric guitar and a small amp over to record. Some clients will also bring gear in.
4. *Power survey*. There are two electrical outlets in his room which are on breaker 4 in the basement. The only other thing on breaker 4 is the lights in the hall. The two outlets are enough for the equipment, but he'll need more power for the air-conditioner.
5. *Air-conditioning survey*. There's space to mount a small window unit, but Iggy will have to get an electrician to run a separate line for it.
6. *Soundproofing survey*. The bedroom is above the dining room in his parent's house. They live in a residential neighborhood in New Jersey. Iggy can't really afford to soundproof, and he doesn't have the space either. So he'll have to keep his monitor volume low and tell his friends with amps not to play too loud. I'll have Joe the carpenter build some soundproofing shutters for Iggy's window and a baffle to fit over his air-conditioner. I'll also have Joe reinforce Iggy's door and add airseals to it. Iggy is a rock musician, but some of his clients do rap and hip-hop. Iggy will have to politely explain to them that he's working in his parent's home and so can't listen at the levels they're used to hearing when they go to clubs.
7. *Acoustical survey*. Iggy's room is a rectangle, and he doesn't have space or money to break up the parallelism. So I'll put his speakers on wall-hung swivel mounts, to keep them away from the surfaces of the walls. Iggy will depend on his near-field monitors to work around the room acoustic. I'll also have Joe pad Iggy's ceiling with crushed fiberglass to dampen the high frequencies a bit.
8. *Architectural survey*. The house isn't very suitable for putting in a professional studio. Iggy will work here for a couple of years and then find a commercial site that's better, once he's developed a client base.

That wasn't too hard, was it? Our example was a home, semi-pro studio, but the same principles of information gathering apply to any size installation. The difference is that instead of a one-paragraph description, the information in each category can run into many pages.

As a final example of Iggy's imaginary installation, a copy of the scale drawing I had him do to find the correct placement for all his equipment is included (Figure 1.39). If I can get an imaginary person to do this kind of work, maybe I can persuade you to do it too.

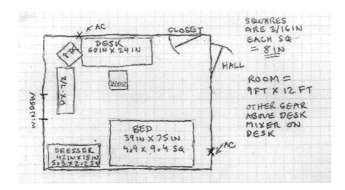

Figure 1.39 Iggy's scale drawing.

Iggy was a good boy, he measured, and drew in, all the necessary details. I hope you follow his example.

'Now are we ready to wire?'

Well, actually, no. If you're really building a studio, you need to go through the information list and fill it out completely. Write down your answers in your work log. Keeping organized notes of your work is critical. Notes can be crude (like Iggy's drawing) but they must be accurate.

If you've been good and gathered all the information, you're ready to go on to the next stage, which is figuring out how much wire, how many connectors and other types of material you will need.

Assuming you've got all the information, you can calculate the length of wire needed to go between each piece of equipment. You can do this on your scale drawing or measure it in real life. (An old Rasta proverb says, 'Measure twice, cut once.') Remember, allow enough slack in your measurements for a good *service loop* and the amount of wire that will have to be cut back to connectorize.

What's a service loop? It is a comfortable amount of slack wire at the connection – enough so it can be cross-patched in case one connection is bad.

You need to leave enough wire to insert/remove plugs easily, to plug to neighboring jacks, and also to allow for re-soldering of connectors at least twice. There is no standard for this, since each situation is unique, but you'll soon learn to estimate how much of a service loop to add. Remember, a little too long is a *lot* better than a little too short.

Add up the individual wire lengths for each piece of gear to get a total amount of wire needed. Make allowance for curvature of wire around corners, up from the floor and around obstructions.

Not all the wire you need will be of the same type. My theoretical eight-track studio might use eight-pair Mogami wire to and from the eight-track, but other kinds of wire elsewhere in the system. The mixer to patch bay wiring might be best done with 24-pair Mogami. The individual effects units might be best done with single-pair wire, like West Penn 291. And I might choose to wire the speakers with Monster cable.

Once you have all your measurements and you've decided which types of wire you are going to use, you're ready to draw up a materials purchase list. If you're not sure what type of wire you'll need, don't worry – we'll talk about all the different kinds of wire at greater length later. The assumption I'm making is that you *will* read the *entire* book before wiring your studio. That's not too much to ask, is it?

Right now, we're talking about what your next step would be if you knew exactly what you wanted to build. You're probably not sure yet of what you want to wire and/or construct. But you can start the materials list in a general way and fill in the details later.

Revisions are a part of any job, so set up your lists with plenty of blank space and skipped lines for changes and writing in details like sources.

Or, do what I do, and slam the whole thing into a computer, in a spreadsheet like Excel, where it's easier to update. And by the way – can you spell *back up*? Regularly. On CDs or flash drives or whatever they've invented lately.

There'll be a whole section on using spreadsheets later; this is still the basic information zone, OK? And handwritten lists are often the raw information you need to make a great spreadsheet that clarifies the job. Don't be afraid to gather data any (and every) way you can.

Include all the different types of wire in the materials list, and be sure to buy 15% more than you think you'll need. You can always cut it back, but splicing in the middle of a harness is difficult.

The same applies for connectors. Add up all the connectors you need, in all their different types, and then add at least four more of each small connector type – maybe three more of expensive multi-pin connectors. This is especially true for those of you who live far away from major supply centers. If you're buying materials mail-order and have to wait for three connectors to finish the job, you won't be happy.

We've talked enough (for now) about information you need to find out and organize. Let's move on to some basic wiring techniques.

Stripping wire

There are several different types of wire you're likely to encounter, and a variety of methods for stripping wire. I'm going to describe each in some detail, because the methods needed depend on the type of wire you're working with.

I talked about these kinds of wire in the beginning, now I want to go into more depth. This approach of multiple exposures to information will make it easier to understand, and retain, what I show you.

One of the most common types of wire used in studios has a *mylar foil shield* around the inner conductors. The mylar is blue on one side and silver on the other. The blue side is insulated (that is, it does *not* pass electricity) and the silver side is conductive (it *does* pass electricity).

The silver side is wrapped around the inner conductors of the wire, including the drain conductor – which has no insulation. Since the silver side of the foil is conductive, it makes excellent contact with the drain conductor, thus providing an effective electrostatic shield for the inner conductors.

Look familiar? The example in Figure 1.40 is of a typical single-pair wire, but foil is also used as an outer shield around multi-pair wire. The foil does a good job as a shield, but it's fragile – if you flex the wire often, the metal on the foil breaks down and the wire becomes noisy and microphonic.

Figure 1.40 Foil-shielded wire.

So *foil-shielded wire is best suited to fixed parts of your installation* rather than use in microphone cables and other wires which are constantly being moved and flexed.

More effective for mic cables is spiral strand shield wire – wire that has bare strands wrapped around the inner conductors (Figure 1.41). This style of shielding is also used by some wire manufacturers (Gotham and Mogami) for multi-pair cables. Spiral strand shield wire is more expensive, but also more effective and durable.

Figure 1.41   Spiral strand shielded wire.

Figure 1.42   Braided strand shielded wire.

Braided strand shielded wire (Figure 1.42) is an older type of wire construction; it's commonly seen in elderly mic cables, guitar cords or cables for vintage condenser microphones. While it's strong and has a good shield, it's a royal pain to deal with, as it must be carefully unbraided a little at a time in order to be properly connected. The best way to unbraid it is to use a small pointy object like a dry-wall screw or a sharply pointed nail, or even the awl tool on your Swiss army knife.

One of the few advantages to this type of wire is that it will hang straight down, without twisting – useful if you're hanging microphones from a tall ceiling and for installations in concert halls, where mics are hung permanently. Otherwise, avoid this type of wire – it's so tedious to work with that you'll lose a lot of time in your wiring.

Figure 1.43 Multi-pair wire.

Multi-pair wire (Figure 1.43) may have either a foil or stranded shield for each pair. Some types have a secondary foil layer wrapped around all the pairs for additional noise shielding. Multi-pair wire is commonly used:

1. Where a large number of wire runs are going along the same path, as from a console patch bay to an outboard equipment rack.
2. For multi-track recorders going to/from consoles.
3. For long microphone snakes going to stage boxes.

No matter what type of wire you're working with, the goal is always the same: to strip off the outer insulating jacket without harming the delicate insulation of the inner conductors. There are a couple of ways to go about this, so let's talk about each one.

Miller-type wire strippers can be used to remove a short length of the outer jacket. This is best done by adjusting the depth-of-cut on the wire strippers to go *almost* through the outer jacket. Then grasp the wire with the jaws of the strippers at the cutaway point, clamp down the jaws on the wire and use a rotating, rocking motion to chew most of the way through the outer insulation jacket. If you've done this correctly, you now have a deep groove in the outer jacket, but you have not cut so deeply as to harm the inner conductors.

Now release the jaws of the stripper and move them slightly toward the end of the wire. Clamp down again, a little less tightly, and bend the wire back and forth at the cut you just made; then finish breaking away the outer jacket. Finally, pull firmly with the stripper's jaws toward the end of the wire, to pull away the section of outer jacket you want to remove.

Caution here: The 'intuitive' thing to do is to make the rotating cut and then pull, taking off the outer jacket at the wire's end. But I'm asking you to do this in two steps. Why not just pull with the jaws of the stripper inside the cut made in the outer jacket? Because the stripper jaws could then gouge the insulation of the inner conductors. My way is a little slower, but a lot safer! Practice this several times with a 1-inch (2.54 cm) strip-back each time.

If you've done this operation correctly, you should now have an exposed section of the inner shield, with no nicks or gouges in it. The examples in Figures 1.44, 1.45 and 1.46 show this being done to wire with a mylar foil shield, but the procedure is the same for wire with a spiral or braided shield.

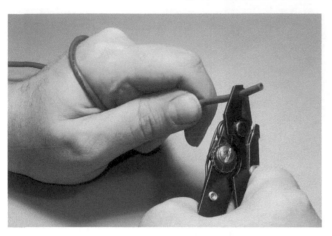

Figure 1.44   Cutting outer jacket.

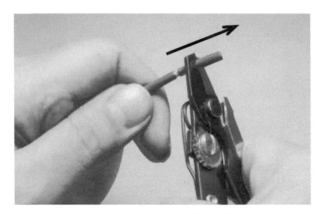

Figure 1.45   Removing cut section – 1.

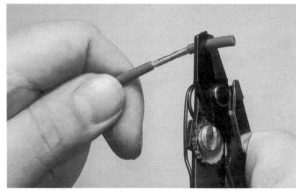

Figure 1.46   Removing cut section – 2.

There's an easier and faster way to do this strip off of the outer jacket – you can combine the small vise you bought with a single-edge razor blade to create a highly effective stripping tool. Mount the razor blade in the vise so that a very small amount of the blade's sharp edge is exposed along the top of the vise. Typically you'll leave about {1/64} inch (0.0397 cm) to roughly {1/32} inch (0.0794 cm) of the blade exposed.

Note in Figure 1.47 that I've taped the end of my finger so I can touch the edge of the blade safely, without cutting myself. I can now safely and repeatably adjust the height of the blade until it's correct for the particular type of wire I'm stripping. Please **be careful** when creating and using these special tools. Razor blades will cut *you* even better than they will cut insulation, so use them carefully! Further, you use them at your own risk. If you are injured, killed or bleed all over your equipment, the authors and publisher will disavow your actions.

Figure 1.47   Mounting blade in vise.

Press the wire down onto the exposed edge of the razor blade and carefully roll it back and forth (Figure 1.48). Keep the downward pressure constant and keep your fingers away from the edge of the blade! Also, keep the wire exactly at a 90-degree angle to the edge of the blade, so as to avoid making a spiraling cut. This will (if properly done) create a perfectly smooth and accurate cut in the outer jacket – far cleaner than is possible with any other method!

Figure 1.48   Cutting outer jacket with blade in vise.

Figure 1.49   Removing cut section – 1.

Figure 1.50   Removing cut section – 2.

Now pull the wire (and your fingers) away from the blade/vise combination (Figure 1.49). Flex the wire at the cut point to finish breaking away the outer jacket. Finally, pull the cut section off the end of the wire with your fingers (Figure 1.50). If the depth-of-cut on the razor blade is correct, this can be done with minimal effort. With a little practice you can do this operation three or four times faster – and far more neatly – than someone stripping the outer jacket off with regular wire strippers.

Once the outer jacket is removed from the end of the wire, you will see either the mylar foil shield, a spiral wrap stranded shield or a braided strand shield.

With mylar foil, to cut away the foil pull down on the outer jacket toward the middle of the wire to expose a little more of the foil than is now visible – say, {1/8} inch (0.317 cm) more exposed foil. Holding the outer jacket in place with one hand, nick the foil with a pair of wire cutters or another razor blade (Figure 1.51).

Figure 1.51   Nicking foil shield.

Figure 1.52   Tearing off mylar foil.

Now tear away the foil at the point you nicked (Figure 1.52). If done correctly, the foil will come away cleanly and the outer jacket will push slightly back toward the end of the wire, covering the point at which you removed the foil. This will provide insulation and strain relief inside the connector you will later attach to the wire.

With a spiral strand shield, the operation is essentially the same as with mylar foil, but you won't have any foil to remove. Instead, pull on the shield strands and lead them away from (at right angles to) the inner conductors (Figure 1.53). Don't twist them together yet; they may be more manageable in an untwisted state. Again, pull down on the outer jacket slightly and pull the shield strands together a little below the cut-off point on the outer jacket. This will keep the shield strands nicely bunched together for you to work on them later. Cut away any strands of insulation exposed during this operation with a small pair of wire cutters (dykes).

Figure 1.53 Pulling shield strands.

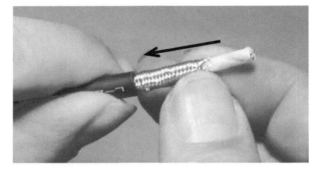

Figure 1.54 Pushing down braided shield.

Should you be unfortunate enough to have to deal with braided shield strand wire, you must carefully and gently unbraid the shield, three or four strands at a time. Once you have all the shield strands unbraided, down to the cut-off point of the outer jacket, gather them together and lead them away from the inner conductors. This is shown in Figures 1.54, 1.55 and 1.56.

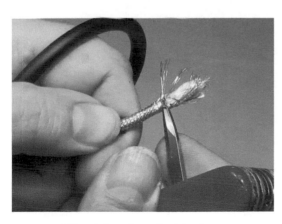

Figure 1.55 Unbraiding shield – 1.

Figure 1.56 Unbraiding shield – 2.

Hold the outer jacket in one hand and push down on the braided shield with the fingers of the other hand, forcing it to compress toward the cut-off point. This loosens the shield and makes it easier to unbraid.

Gently and carefully, slowly unbraid the shield conductors. Try *very* hard not to lose any in the process. Keep going, you're almost done! Unbraid the shield all the way down to the breakout point (where the outer jacket was cut away).

You'll then be left with a lot of fuzzy, stringy insulation to get in the way of your soldering. Carefully cut it all away with a small pair of wire cutters. Try to remove all the insulation – it can foul a solder joint you create and make it unreliable.

Carefully separate the wire strands from the fuzzy, frizzy packing strands (the non-electrical stuff; Figure 1.57).

You may, or may not, be able to pull down on the outer jacket, depending on the type, age and condition of the wire. If possible, pull the outer jacket down and unbraid a little below the outer jacket's cut-off point. If you can't pull it down, just unbraid right down to the cut-off point. Again, your goal in unbraiding the shield is to lose as few of the strands as possible – ideally, none.

Figure 1.57   Separating strands.

After tucking the inner conductors and shield out of harm's way, carefully cut away all the exposed insulation (Figure 1.58). Make absolutely *sure* that you're not cutting the conductors/shield, or you'll get to do the whole strip-out over again with a slightly shorter wire.

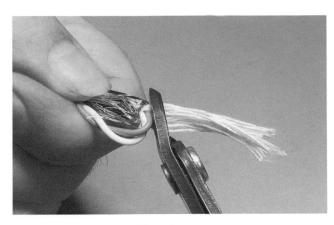

Figure 1.58   Cutting off insulation.

OK, you've practiced stripping the outer jacket off some scrap wire. You've done it several times – as many as it took to learn to do it properly. What next?

The next step is to strip back the inner conductors of the wire a small amount. Typically, an {1/8} inch or less strip-back will give you enough of the exposed strands to tin and solder them properly. One exception to this is old-style RCA male plugs, which must be stripped longer. RCA connectors are covered in their own section, so I won't digress here.

I recommend using a pair of wire strippers for cutting the insulation on the inner conductors. If you've adjusted your wire strippers for the outer jacket insulation, you'll need to readjust them to the correct size for the insulation on the inner conductors. This can get pretty tedious, so I recommend having two pairs of wire strippers. Set one pair for your outer jacket thickness and the second pair for the insulation of the inner conductors.

Remember that mylar foil shield wire I first showed you? I'm going to go back to it now and we'll ignore braided shield wire for a bit. It's way too slow to work with for large numbers of connections. Foil shielded wire is the most common type for fixed installations, so I'm going to show you the next techniques on that type of wire.

Our first step is to cut the inner conductors, and drain wire, to a suitable length for the connector(s) you're wiring. Our example here is for an XLR, so they can be pretty short. I'll tell you just how short at the end of the sequence (no peeking ahead now).

Watch what you're doing, and cut all three conductors to the right length (Figure 1.59). Mistakes here are irrevocable except by redoing the whole process.

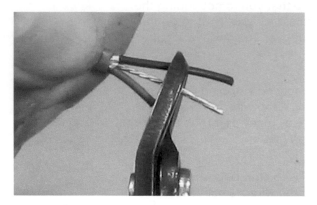

Figure 1.59 Cutting inner conductors.

Set a pair of strippers for the inner conductors and strip back a small amount of insulation, about {3/32} inch (Figure 1.60). The insulation will 'wick back' when the strands are soldered, exposing more of the strands than you've stripped.

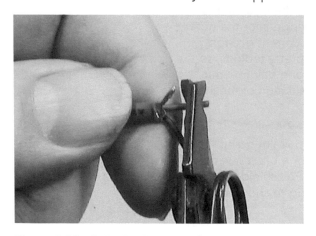

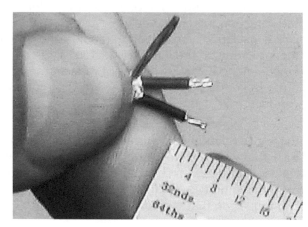

Figure 1.60 Stripping inner conductors.   **Figure 1.61** Finished length (for XLR).

The finished strip-off is shown in Figure 1.61. Well, almost – there's that little bit of mylar foil to cut away. This is a typical length for an XLR plug – either male or female. The exposed strands of the inner conductors are about {3/32} inch and the insulated length of the inner conductors is around {7/32} inch.

For a physically strong connection, your tolerances on this connector should not vary more than {1/16} inch. Oodles of room!

What you've just seen and read gives you the rudiments of stripping wire. Now it's up to you to use this information and *practice* it. Take some scrap wire and practice stripping it back a few times. You are ready to move on when your stripped back wires are clean and accurate, with no nicks or gouges. If you need to practice more, please do so now. Reading and talking about an action are no substitute for actually doing it!

You don't learn to ride a bicycle by reading about riding bicycles. You learn to ride by getting on the bicycle and falling down, picking yourself up and trying again. Similarly, you will only learn to be a good wireperson by doing it – there simply is no other way.

If you've followed instructions, and practiced your new skills a bit on some scrap wire, you're now ready to learn the next steps – dealing with multi-pair and tinning wire.

Tinning wire is the process of coating it with a layer of molten solder, which also flows in between the copper strands of the conductors. It makes the wires easier to solder and also makes for a better connection. The process name comes from the fact that the basic alloy of solder is a tin/lead combination. This alloy is improved by the addition of a small amount of silver.

Normally, you would only tin wire when you are ready to attach it to the connector. But just as you practiced stripping wire, it's important for you to practice tinning wire.

Getting impatient? Take a deep breath. Drop your shoulders. The time you take to read and practice now will be paid back many times over in your actual installation. Not only will you be faster, but the quality of your workmanship will be much better.

Dealing with multi-pair wire

All of the instructions up to now have dealt with putting one connector on one wire – which is the most basic form of wiring. But what if you're dealing with multi-pair wire which has many discrete wires bundled together in an outer jacket? In that case, you might have up to 32 individual wires inside an outer jacket.

Figure 1.62 Back to multi-pair wire.

To work on multi-pair wire (Figure 1.62), you must first remove a suitable length of the outer jacket *without* harming the insulation of the inner wires. This is almost impossible with most commercial wire strippers, but is done easily with a little practice and the special 'home-made jacket slitter' explained on page 19 of this section. If you haven't already made the jacket slitter, please do so now, before you continue.

After you've made your own home-brew jacket slitter, you're ready to use it on some multi-pair wire. Make sure the amount of exposed blade is small enough to not harm the insulation of the inner wires. Typically, an exposed edge of {1/64} inch (0.0395 cm) to roughly {1/32} inch (0.079 cm) of the blade should be left visible.

Practice using the jacket slitter on some scrap wire *before* you try it on your multi-pair wire – this is the *only* way you'll avoid mistakes and damage to your new wire!

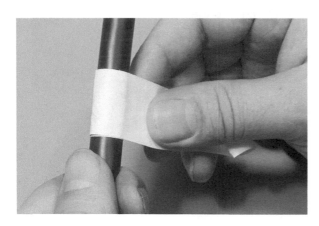

Figure 1.63 Taping wire at cut-off point.

The next step is to figure out how far back to strip off the outer jacket of the multi-pair wire. Typically, this will be 1–3 feet (30.48–91.44 cm) but will vary greatly with your particular installation.

Measure off the amount you want to strip back the outer jacket and put a couple of turns of artist's tape or masking tape around the outer jacket of the multi-pair just past the point you've measured (Figure 1.63). This tape will act as a guide when you score the outer jacket with the jacket slitter. Make sure the tape is exactly at a right angle to the wire.

Grasp the jacket slitter between thumb and forefinger in such a way that your fingers will slide along the wire and act as a guide for the slitter (Figure 1.64). Note that the slitter is at an angle to the work's surface. By varying the angle, and the force of your cut, you can adapt perfectly to different types of wire with different amounts of outer jacket thickness (Figure 1.65).

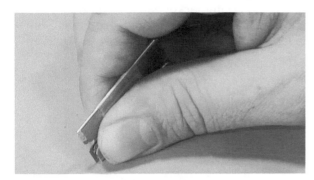

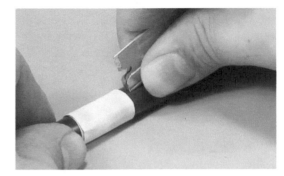

Figure 1.64   Correct grip of slitter.   **Figure 1.65**   Starting cut of outer jacket.

Run your fingers and the slitter lightly along the wire from your marked point in a straight line to the end of the wire – in other words, parallel to the wire, from the mark to the end of the wire. Bear down more on the blade at the very end of the wire – this will help you start the tear-off of the outer jacket.

Keep working down the wire with the same pressure and angle of cut (Figures 1.66 and 1.67). The ideal is that the outer jacket is 70–90% cut through, without the slitter ever actually touching the shield or inner conductors. I can do it every time. You can too, if you concentrate!

Figure 1.66   Continuing cut with slitter – 1.   **Figure 1.67**   Continuing cut with slitter – 2.

Don't stop in the middle for a smoke. Don't vary the pressure or the angle of your cut. Bear down hard at the last {1/2} inch to be cut. This will make the outer jacket easy to tear at the end of the wire.

Still grasping the slitter properly, make a circular cut around the outer jacket just above the tape you placed on the outer jacket earlier (Figure 1.68). Make sure the ends of your circular cut match up and mate, so the cut-off is always at a right angle to the length of the wire. Use the tape as a guide to help you do this.

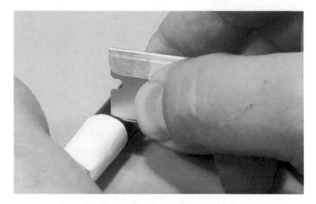

Figure 1.68   Rotary cut at breakout.

Now go to the end of the wire and break open the insulation at the scored point with a pair of dykes (wire cutters). If you've done all the above operations correctly, the outer jacket of the wire can now be torn (peeled) off the bundle of wires by hand. Peel it all the way back to the cut-off point you marked with tape (Figure 1.69).

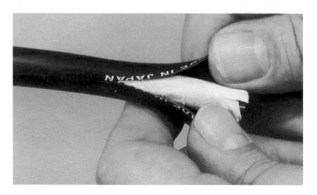

Figure 1.69   Peeling back outer jacket.

Figure 1.70   Tearing outer jacket at cut-off point.

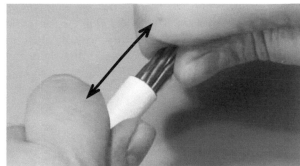

Figure 1.71   Pulling back outer jacket.

After you've peeled back the outer jacket to your cut-off point at the tape (along with the circular cut you made there), cut gently along the circular cut you already made, so you can tear away the outer jacket with your hands. Then tear the jacket away (Figure 1.70).

Pull back hard on the remaining outer jacket to expose more of the bundled wires underneath (Figure 1.71). This extra pullback adds strength to any strain relief attached to the outer jacket of the wire. Cut away any paper (or other) insulation exposed by removing the outer jacket.

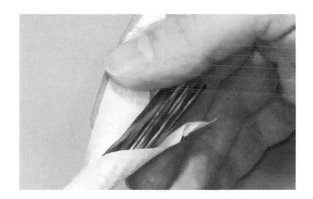

Figure 1.72   Removing bundle wrapping.

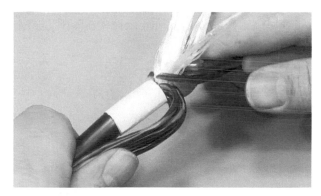

Figure 1.73   Cutting away insulation/wrapping.

Typically, the first thing you'll see after removing the outer jacket of multi-pair is some kind of wrapping around the wires inside (Figure 1.72). It may be paper or plastic, but it was put there to help form the wire when it was made. We don't need it, so cut it ruthlessly (but carefully) away (Figure 1.73)!

Take the wires you've now exposed and unspiral (untangle) them, so they lie as flat as possible (Figure 1.74). The ends of the wires will be different lengths now, because they were wound around each other in a particular way when the wire was formed.

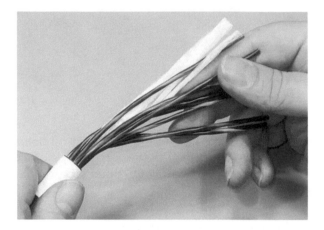

Figure 1.74   Untangling wires of the multi-pair.

Now use a pair of wire cutters (dykes) and cut back the ends of the individual wires, so they are all equal (or very close to equal) in length. This process is called 'justifying' the wires (Figures 1.75 and 1.76).

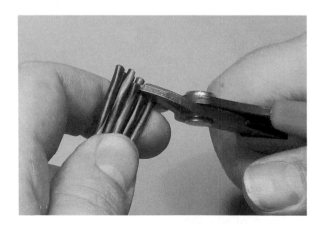

Figure 1.75   Equalizing (justifying) length of wires.

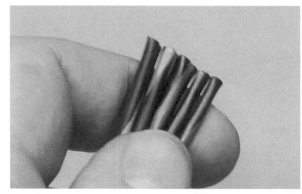

Figure 1.76   Fully justified wires.

Congratulations! Your multi-pair wire is now ready to be connectorized or made into a snake. One last caution – the insulation of the inner wires in multi-pair is very thin and the wires themselves are small and fragile. Be nice to them.

For this reason, it's sometimes helpful to add a cap-off of heat-shrink at the end of the wire, so strain reliefs have something to grab onto. This is true even when there is no other heat-shrink used in the wiring of a connector, and especially true with thin multi-pair wire.

Figure 1.77 shows our old friend from the heat-shrink discussion. Even without the green (drain conductor) heat-shrink, the wire cap-off shrink adds strength and thickness for strain reliefs to 'bite' on.

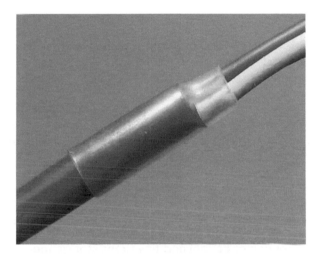

Figure 1.77   Heat-shrink cap-off on wire.

Now that I've shown you how to strip both single- and multi-pair wire, it's time to talk about the next step in preparing your wire for soldering and connectorization – tinning.

Tinning wire

Take the time now to set up a practice workstation. Light it properly, provide for adequate ventilation and a good work surface. Run some AC power over to it, put your soldering iron in its holder, place your tools within easy reach and your vise in front of you.

Caution: get a pair of safety glasses and wear them at all times when soldering.

You'll also need some solder to practice with, along with that bit of scrap wire. I recommend using 62/36/2 solder instead of the standard 60/40 blend. The numbers refer to the percentage of tin/lead/silver in the solder alloy. Solder that is 60/40 or 63/37 has no silver in it, only tin/lead. I've found that 62/36/2 solder makes better connections and is easily available from Radio Shack. The catalog number is 64-013 (URL: http://www.radioshack.com/product.asp?cata log%5Fname=CTLG&category%5Fname=CTLG%5F011%5F009%5F007%5F 001&product%5Fid=64%2D013&FGL=1-001).

If you can't get solder with 1–2% of silver in the alloy, try to get a 63/37 blend of tin/lead. Even that slight change in the solder alloy will make for better connections.

Caution: be sure to use rosin core solder rather than acid core solder. Acid core solder is designed for plumbing work and will destroy any electronic equipment that you use it on! Wires are no exception.

Solder comes in different thicknesses, just like guitar strings. I like to use a medium-gauge solder, around 0.050 inch (0.127 cm) for soldering connectors. The really thin types of solder are designed for finer work, like soldering components to a printed circuit board (PCB).

Just as you can get a better or worse tone for playing specific music with a certain size guitar string, you'll get better or worse results soldering specific components with different sizes of solder.

Medium-sized (connector gauge) and tiny (transistor gauge) solders are illustrated in Figure 1.78. The thin transistor gauge solder is really hard to see! A close-up is shown in Figure 1.79. That's better! The thin solder is about the size of heavy sewing thread, so the big stuff isn't really all that huge. The actual diameters are 0.050 inch for the medium gauge and 0.015 inch for the tiny stuff. Yikes, that's small!

A couple of useful tips:

1. If you only have very thin solder, and are wiring heavier work, take the solder and double a length of it over itself one or more times. Then twist the solder together and hey presto – thicker solder that suits your work.
2. If you have solder that's too thick for your work, apply it in tiny dabs to the components being heated.

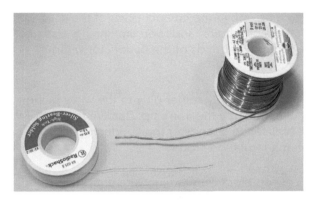

Figure 1.78 Two sizes of solder.

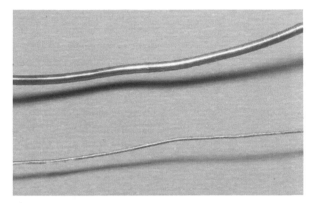

Figure 1.79 Solder close-up.

Time to practice tinning

'Tinning' means putting a little solder on the tip of the iron and allowing it to melt, so that molten solder will flow more easily. That's tinning the tip of your soldering iron.

However, like so many wiring terms, tinning has more than one meaning. Tinning wire is the process of coating it with a layer of molten solder, which also flows in between the copper strands of the conductors. Doing this makes the wires easier to solder and also makes a better electrical connection.

Remember when you practiced stripping the inner conductors on a wire? Take that wire and mount it in the vise. Plug in your soldering iron and allow the tip to come up to heat. Put some solder on the tip of the iron and allow it to melt. You *do* have those safety glasses on, right?

Wipe off any excess solder or rosin with your soldering sponge, but allow a small amount to remain on the tip of the iron. The fresh solder on the tip of the iron provides a 'wetting' action so that molten solder can flow easily from the iron tip onto the work being soldered.

Place the iron's tip on the exposed strands of the inner conductors and hold it there a few seconds to allow the metal of the conductors to come up to heat. Apply a small quantity of solder at the point of contact between the iron and the strands of the conductor.

Once the solder has begun to melt, move the contact point of the solid solder and the soldering iron tip along the strands you are tinning. Use just enough solder to flow evenly onto the strands of the conductor without leaving blobs of solder.

Some people will say that you should apply solder only to the wires being heated and not at the point of contact by the iron's tip. But I've found this technique less satisfactory, because the strands of the conductor become overheated and tend to melt the insulation jacket.

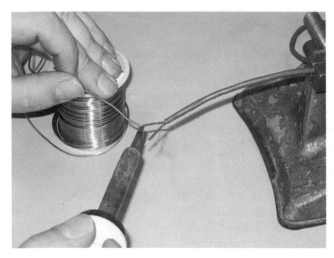

Keep practicing tinning wire until you can produce a smooth, even coating of solder on the strands of the inner conductors without excessive melt-back of each conductor's insulating jacket (Figure 1.80).

Some types of wire use insulation with very poor thermal stability, so the melt-back or 'wicking' will seem excessive, no matter how careful you are. When working with this type of wire, you must allow extra length in your strip-back for the strands you will cut away after they are tinned.

If you need to strip some more scrap wire to practice on, please do so now.

Figure 1.80   Tinning inner conductors.

Tinning (filling) solder cups

Good, we've got the wire ready and you know how to deal with it. Now let's prepare a typical XLR connector for soldering.

There are two ways to approach this situation. One is the 'old-school' method of pressing the conductors into the solder cups with one hand, feeding the solder in with another hand, and using another hand to hold the soldering iron. If you've been counting, that's three hands, which is hard for most folks.

Now you could use the Third Hand vise I showed you, or two small regular vises. Even tape the wires onto something to position them while soldering.

I've done all those things, and been happy I was able to do them. But there's an easier way.

Suppose you could load each solder cup with just the right amount of solder. And your solder was of high enough quality to take reheating (once) without degrading its conductivity or strength. Often, that's what production-line solder people do: pre-load the connectors with solder, tin the wires, and then quickly reheat the loaded solder to insert the conductor's strands into the solder cup.

I'm now going to give you a brief look at this 'secret' procedure for connectorizing. Later, in our section on actual soldering, we'll revisit this technique. Right now, I just want to give you an overview, the big picture, before we dive into itty-bitty details.

Place the iron's tip against the solder cup in a way that maximizes the contact between the two (Figure 1.81). This will heat the cup in a few seconds. Then feed the solder into the cup, allowing time for it to melt, and flow. Sometimes an air bubble can get trapped in the cup; if that happens, allow the bubble to burst, then add a little more solder to fill the cup.

Figure 1.81   Filling solder cups of XLR.

The ideal is that each cup is full enough to surround the strands inserted into it, but not so full as to overflow when the strands are pushed home. Think of a rounded teaspoon of sugar – rounded, but not heaped, if you see my point.

In Figure 1.81 only one of the three cups being filled is shown, but of course for this technique to work, all three have to be prepped (that's wirespeak shorthand for 'prepared'). Got those solder cups tastefully filled? Nice bright, shiny solder, with a bit of rosin left on for the reheat? Good! Now let's attach a conductor to a solder cup.

Inserting conductor strands

No, that's not a suggestive title, it's just what you're doing at this step – honestly. Place the iron tip across the cup to heat both the cup and the solder. Wait one or two seconds after you see the solder melt and start to flow. This allows all of the solder in the cup to become molten. Then quickly push the strands into the cup (Figure 1.82). Push them in until the conductor's insulation is flush to the rim of the cup. Rapidly slide the iron away with a sideways motion (don't lift it).

Figure 1.82   Inserting strands into hot cup.

If you've done all this correctly, and held the conductor in place for a few seconds while the solder cooled, you should be rewarded by a chrome fender bright solder joint, with excellent strength and conductivity.

In Figure 1.82 only one conductor is shown being soldered, when in reality all three must be soldered. But you knew that, you were just waiting to see if I'd forget to tell you, right?

Congratulations! If you've got this far and practiced the techniques I've described and illustrated, you're now almost ready to begin wiring.

Designate a few connectors of each type for practice, and work with them before actually starting the wiring of your installation. This will save you some ruined connectors and a lot of frustration.

The wiring techniques for each individual type of connector are contained in later sections in this book. Select the sections that contain the types of connectors you'll be working with and study each one carefully.

But, before you rush off to start wiring things, *please* check out Section 2 of the AWG. That's the Intermediate Information section, and there are a few things I'd like to show you there, before you start chopping.

The last technique I want to talk about in this section is that of de-soldering. If you practice on a few connectors, it would be useful (and cheaper) if you could do so several times on each connector. And suppose you make a mistake on a 'real' connector? So I'm going to show you a rapid and effective way to de-solder connectors. Remember those safety glasses I told you to purchase? Now is a great time to put them on, if you're not still wearing them.

Now let's de-solder

After you remove a conductor from a connector, a thin coating of solder will always remain on the conductor. In addition, the majority of the old solder will remain in the solder cup and must be removed.

To de-solder is to take away the bulk of the old solder by a variety of means. The goal is to remove the majority of the old solder, without damage to either the insulation of the conductors or to the connector itself. This is necessary because reheated/contaminated solder makes a poor electrical connection, and is also physically weak.

It's OK to leave a light film of solder on both the conductor and the solder cup. This actually aids the re-soldering process and is, in any case, impossible to remove.

Let's use an XLR male plug as our first example. It has some conductors attached to it which we want to remove along with the old solder in the solder cups.

Mount the connector (after removing its shell) in the vise. Tighten the vise jaws on the Pin 3 contact of the plug or on the plastic surrounding all the pins. If you tighten the vise jaws across the Pin 1 and 2 contacts of the plug, they may become loose or misaligned when you start to de-solder the conductors. The plastic in which the contact pins are embedded is somewhat thermoplastic in nature and will flow when heated. Then the pins will become loose or misaligned – not a good thing.

Grasp the most convenient to you of the conductors with the tip of a pair of small needle-nose pliers. While holding the conductor with the pliers, heat the solder cup the conductor runs into. Pull the conductor out of the solder cup once the solder has become molten. Perform the same actions for the other two conductors.

In Figure 1.83 Ken's camera got me just as I'd pulled the conductor out of the center solder cup (Pin 3). And yes, you have to do them all.

Figure 1.83   Removing soldered conductors.

Now you have a plug that's free of conductors, but you still have the old solder in the solder cups. It's been used and contaminated; you'd like to get it out, but how?

Remove the connector from the vise and grasp it by the plastic with your pair of slip-joint pliers (Figure 1.84). Hold the connector so that the solder cups face toward you. Heat any of the three cups with the tip of your soldering iron and, *while the solder is still molten*, slam the pliers down against the edge of the vise! Make sure you hit the vise with the edge of the pliers and not the edge of the connector (Figure 1.85).

Figure 1.84   Heating solder cups.

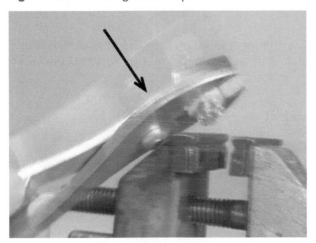

Figure 1.85   Hitting vise to expel solder.

This action will cause the molten solder to fly out of the solder cup and (hopefully) down onto the vise or the tabletop. Keep those safety glasses on – and while I'm on the subject, it's not a good idea to do this while you're wearing shorts. It's far less painful if stray drops of solder hit the fabric of your trousers than if they hit your bare skin.

Repeat this operation for both of the other two solder cups in the connector. The plug will now be quite hot, so let it cool for a few minutes before you attempt to handle it.

De-soldering other types of connectors is similar, but the exact details will vary from connector to connector.

You can also use a solder sucker or solder wick to remove the solder from the cups. But what if you run out or can't buy any? The above technique works every time, and all you need are the tools you should already have.

This method of de-soldering is pretty universal. You could call it the PHW method for: Pull out the conductor, Heat the solder cup and Whack the vise with your pliers holding the heated plug, to expel the solder.

A number of people asked that we show how to apply it to a {1/4} inch guitar plug, since they are so ubiquitous. Here is the same sequence of operations for de-soldering a {1/4} inch male guitar plug. The only changes are to accommodate the difference in construction of guitar vs. XLR plugs.

Unscrew the outer barrel and mount the plug in your vise with the bent-over arms of the strain relief facing up. In Figure 1.86 I'm prying up the first arm with a small screwdriver.

Figure 1.86   Mount plug and open strain relief.

Once I have the edge lifted up, I'll grip it with pliers and spread it wider. Then I'll do the same operation on the other strain relief arm. I'm done when both arms are spread widely enough to let the wire just fall out when the conductors are de-soldered.

As the strain relief has already been stressed by bending, I want to do this very gently.

I'm using a stereo guitar plug as a more complicated example, so I'll pick one of the conductors, heat the solder tab it's attached to, and pull the conductor away as soon as the solder melts (Figure 1.87). Then I'll do the same action for the other conductor.

Figure 1.87   Removing soldered conductors.

Figure 1.88   De-soldering shield.

This still leaves the shield of the wire firmly attached to the strain relief arm – but that's our next step (Figure 1.88).

Figure 1.90   Hitting vise to expel solder.

Figure 1.89   Heating solder tabs.

I've rotated the plug in the vise, so that the strain relief arm and the attached shield are facing up. Now to heat the shield, until I can pull it loose. With both conductors and the shield detached, the wire will simply fall away.

It's a good idea to slide the iron tip back and forth on the strain relief arm, to spread the leftover solder into a thin, smooth film. Any excess can be reheated and struck off when we do the same action for the solder tabs (Figure 1.89).

Just as for the XLR, I'm holding the connector in a pair of pliers, heating each tab up to melt point, and then hitting the vise, as shown in Figure 1.90, to strike off any excess solder. For a stereo guitar plug, I'll do this twice for the solder tabs and a third time (if needed) for any excess solder on the strain relief arm itself.

This action will cause the molten solder to fly off the solder tab and (hopefully) down onto the vise or the tabletop, just as it did for the XLR. The solder will be just as hot, so safety glasses on, cover your exposed skin, and watch which way you fling it!

As we've seen, the actions needed for the two types of connectors shown are very similar. Use your common sense to extrapolate for other types of connectors.

A word about cleaning things

There's an aspect of both soldering and de-soldering that I haven't talked about – which is cleaning your work and any connectors you recycle. The reason is, this is almost never done in the field – but it *should* be done for every connection.

Commercial electronics manufacturers spend good money to de-flux their connectors and printed circuit boards. They do so because they know that it extends the life and reliability of their equipment.

This is equally true of any field wiring you do – but despite this, I'd guess that more than 99% of field solder work is not de-fluxed and more than 99% of recycled connectors are not cleaned before reuse. However, if you want really pure, audiophile-level connections, you should de-flux your work. There are commercial de-fluxing liquids, or you can use a series of baths and a small, stiff brush – a toothbrush works well.

One good commercial de-fluxer is made by Caig Laboratories – it's called Flux Wash. More info at: http://store.caig.com/s.nl/it.A/id.2514/.f?sc=&category=1790. Flux Wash works very well – but that toothbrush helps it along.

When I built 24-track analog recorders we would use three baths and brushings – acetone, alcohol and plain water. The only problem was finding a toothbrush that wouldn't melt in the acetone.

Remember that alcohol and acetone are highly flammable, and the fumes aren't good for you either. Even if you use commercial de-fluxing liquids, wear gloves, work with good ventilation, and keep all cleaning agents away from flame, heat guns and hot-tempered individuals. Work with small batches and discard the leftover fluid(s) immediately.

Another aspect of cleaning is that we show work being done with new wire and new connectors. You folks in the field may be dealing with old, corroded wire and dirty, nasty, old connectors.

That film of corrosion/gunk must be removed before a good solder connection can be made. You can scrape it off with a blade, file it gently, use crocus cloth or even a pencil eraser for delicate items.

But if you see corrosion on either your wire or plugs, it must be removed before attempting to solder. The same applies to dirt, grease and even oil from your fingers. If you are not working with new materials, clean the items to be soldered first!

The end of the beginning

This is as far as we can go in the introductory section. You've learned what tools are required, what information must be gathered, and what techniques must be practiced to successfully wire your studio.

We're going to go over some of these topics again in the next section. This was the overview. Too much detail at once may obscure the general concepts you must learn before specific details can become meaningful.

Now it's time to study Section 2 – Intermediate Information, and the connector sections that deal with the specific connectors for your particular installation. With a clear game plan, lucid instructions and vivid pictures to guide you, I'm sure you'll do well when you get down to work. Good luck and happy wiring!

And before you start wiring, please be sure to practice? It's not only the fastest way to Carnegie Hall, it's also the fastest way to a well-working studio!

Disclaimer

All techniques and procedures described in this book are used at the reader's own risk. The author and publisher assume no responsibility for any damage or injury incurred by their use.

Intermediate Information

The good/bad/ugly and the 'bead game'

Some bad examples and a good one

So far, I've only shown you well-done work – it looks the way it's supposed to look. But how will you know what to avoid if I don't show you some examples of bad work? Sloppy, shoddy work that no one will admit to doing, but that somehow occurs all too frequently.

You've seen the good – and there's plenty more of it in Section 3, the modules on specific connectors. Here is a brief 'rogue's gallery' of common mistakes. Or maybe I should say 'come on missed takes'?

After the 'bad guys' I want to talk about a technique called 'beading'. It's described in some of the connector modules, like 3.3 (mini-male guitar plug connectors) and 3.7 (TT male connectors), but if you don't read those modules you might miss it. And beading is an important technique; sometimes it's not possible to add solder while heating a conductor – the strands of the conductor have to carry an extra 'payload' of solder. That's what beading is for.

But back now to our 'usual suspects', the most common errors of bad soldering – the typical villains. And after I show you all the wrong ways, I'll show you a good solder connection to inspire you.

Hehe, sure is too much solder there in Figure 2.1.1, right? That elephantine blob of solder, looking like a large metal goiter, may be exaggerated, but I'm trying to make a point. Use only enough solder to fill the solder cup, allowing for the amount of solder that the wire will displace when it's inserted into the solder cup – not too much solder, not too little.

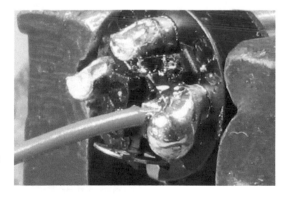

Figure 2.1.1   Too much solder.

Speaking of too little solder, a classic example is seen in Figure 2.1.2. This solder job has another problem too. The conductor has too much exposed metal. The insulation should go right up to the edge of the solder cup.

Figure 2.1.2   Too little solder.

An even more extreme case of 'too much exposed metal' is illustrated in Figure 2.1.3. This weakens the solder joint and can short to adjacent conductors.

Figure 2.1.3   Exposed metal.

In Figure 2.1.4 the insulation has been inserted too deeply in the solder cup. This can cause contamination of the solder joint with the plastic of the insulation, degrading its conductivity. Not too deep, not too far outside the cup either. Remember Goldilocks and the three bowls of porridge? There's a position that's 'just right'.

Figure 2.1.4   Insulation too deep.

See the gray, mottled color/texture of the solder in Figure 2.1.5? That's
a sure sign of a 'cold' solder joint. It is physically weaker and higher in
resistance than a properly done connection. If one of your joints looks like
this, it won't help to add more solder! Instead, you have to take all the
old solder *out* and re-tin the solder cup with fresh solder. Once solder has
been overheated, or re-re-re-reheated, its conductivity and strength are
degraded. It must be replaced.

Figure 2.1.5   Cold solder joint – 1.

A cold solder joint isn't the only problem in Figure 2.1.5. There's too much
exposed metal and the solder distribution in the solder cup is uneven – too
much on the right-hand side and not enough on the left-hand side.

To top it all off, the conductor's been overheated – we can see the insulation beginning to melt back.

Figure 2.1.6 is another shot of our cold solder joint, along with the other ways this connection is troubled. See the gray color compared to the bright silver on the other solder cups?

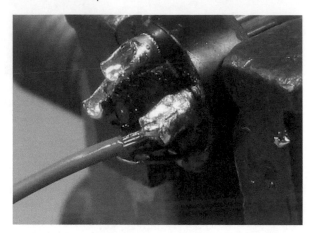

Figure 2.1.6   Cold solder joint – 2.

It's not hard to burn the insulation on conductors (Figure 2.1.7) – just leave your soldering iron in the wrong place for a few seconds. It's a lot harder to fix it once burned, and requires thin diameter heat-shrink, along with a heat-shrink gun.

Figure 2.1.7   Insulation burned.

I hope you never see mistakes like this. I hope even more that you never *make* mistakes like those you've just seen. But if you do, remember, mistakes are part of the learning process. Don't be (too) upset. Don't get angry at yourself. Just do it over until you can do it right – even if that means doing it several times.

I promised you a 'good' example at the start of this section, so now it's time to (re)visit an example that occurs in Section 3 of this book. It's the picture of a properly soldered connection on an XLR plug (Figure 3.6.30). Here, it's been cloned and renamed.

All the aspects of the solder joint are correct in Figure 2.1.8. The solder is bright, shiny and fills the solder cup completely, without being either underfilled or overflowing. The conductor is correctly inserted, with the insulation pushed flush to the edge of the solder cup, without being inserted too far or leaving exposed strands. Isn't it pretty?

Figure 2.1.8   Correct solder joint.

The 'bead game'

With apologies to Hermann Hesse, our bead game is not so metaphysical, nor are the beads made of glass. Rather, the beads are small grape-shaped blobs of solder, deliberately created on the ends of the conductors we want to solder. The function of our inedible grapes is to bring a 'payload' of solder to the solder joint, in situations where it's not practical to do so by hand.

To do this, it is necessary to reheat the solder once but, if you're fast, this can be done without degrading the electrical or physical qualities of the solder. Figure 2.1.9 shows how to do it.

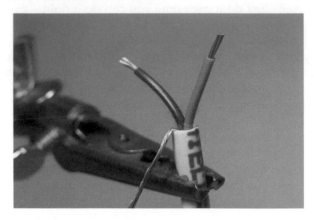

Figure 2.1.9   Ready to tin.

Since this is just an example of beading, I'm only going to tin and bead the red conductor. Here's a pre-stripped wire ready to use. You better already know how to strip wire if you're this far along in the book!

Sometimes you can tin and bead in one operation – it depends on how cooperative the wire is, the type of solder you're using, and the soldering iron you're working with. Often, it's better to do this as a two-step operation – tin first and then bead. In Figure 2.1.10 I'm tinning the red conductor with a thin coat of solder. This will make it easier for my bead/blob to adhere on the strands of the conductor.

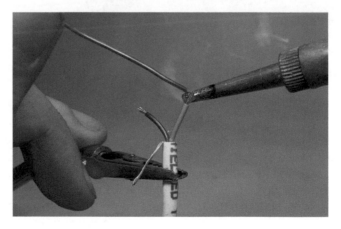

Figure 2.1.10   Tinning conductor.

Figure 2.1.11 shows the finished tinning. Notice the slight mounding of the solder coating? The conductor is ready for the next step – beading.

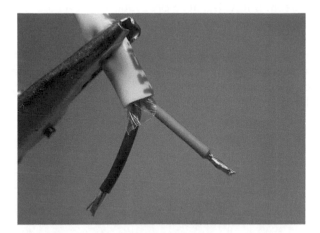

Figure 2.1.11   Tinning done.

Notice that I've inverted the wire, so the molten solder will flow down, toward the tip of the conductor (Figure 2.1.12). It will also flow nicely off the conductor and onto your bare knee if you place it in harm's way.

Figure 2.1.12   Beading conductor.

Sometimes it's necessary to let the bead/blob fall off and quickly try again – so do this on a tabletop, wear safety glasses, and be aware of where your body parts are, if you don't want molten solder burning a crater into your skin!

If you do get hit with hot solder, shake it off; don't touch it or try to rub it off. You'll just embed it deeper into your skin and make the burn worse. Solder in your eye is an immediate trip to the emergency room; don't try to deal with it yourself. But with a little precaution (and a little luck), none of these bad things will ever happen to you, right?

I admit it – the bead in Figure 2.1.13 is not grape shaped. It's more like a frozen water drop, but it's still acceptable for working with. There's enough solder, but not too much. And it's nice and shiny, so it probably wasn't overheated. I'll use it.

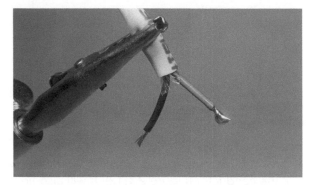

Figure 2.1.13   Finished bead.

A bare RCA male connector I had handy is shown in Figure 2.1.14. We always want to tin the solder cups in connectors, so that's what I'll do with this one to ensure good bonding with the bead.

Figure 2.1.14   Bare connector.

I don't have to put much solder in the cup, since my bead has a lot (Figure 2.1.15). Normally, I'd fill the cup with solder and not bead the wire, but sometimes there's no cup to fill, just a flat tab or, worse, a convex surface, like some TT plugs have.

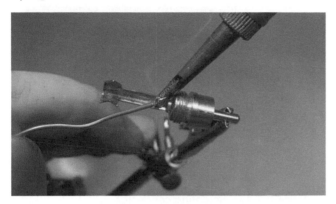

Figure 2.1.15   Tinning solder cup.

This connector is just being used for a demonstration of beading techniques because I had it handy, and didn't have any of the more esoteric connectors that really need to be beaded available. So shoot me. If you want an example of where beading is absolutely mandatory, refer to module 3.7 (TT male connectors).

Oops, there's a bit more solder there than I intended, but it's clean and bright (Figure 2.1.16). Let's try soldering the beaded conductor in place and see how it looks.

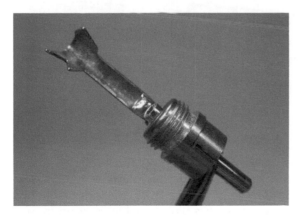

Figure 2.1.16   Ready to connect.

I've placed the bead on the solder cup and pressed the soldering iron down firmly on both, for quick heat transfer (Figure 2.1.17). I'll also push the conductor a bit forward into the cup, when the solder melts.

Figure 2.1.17   Soldering conductor.

The completed solder point is shown in Figure 2.1.18: shiny, with no overflow from the cup, and the insulation goes right up to the solder joint. I admit, it's ever so slightly too plump – I could have added less solder to the cup. But it's good enough to show you the concept of beading.

Figure 2.1.18   Completed connection.

Since this is just an example of beading, I'm not going to deal with wiring the whole plug. If you want to see that, read Section 3.4, the module on wiring RCA male connectors. This particular connector will be de-soldered and put back in my spare parts bin. Recycling is good for electronic parts too.

Using a DVOM or DMM, and what is it anyhow?

Way back at the beginning of the book, I promised to show you how to test the cables you create. To do this, we'd typically use a gadget called a 'digital volt-ohm-milliammeter'. Since saying that whole name repeatedly quickly becomes tedious, we use the acronym DVOM. It is also called a DMM, for digital multi-meter. Both names are good, describing the unit accurately. The older version of this meter, with the swinging needle, was called a VOM, as it's not digital.

I'll show you one of these shortly, and also give a detailed description of some of its typical functions. But before I get so specific, I'd like to explain why you might want to use a DVOM to help you in your work.

All DVOMs measure electricity in at least three ways. First, they measure voltage: DC (direct current) voltage and AC (alternating current) voltage. As you may have guessed, voltage is measured in volts, multiples of volts and fractions of volts.

Second, they measure DC resistance, which uses the ohm as a basic unit of measurement. The same principle applies here: we can have fractions of an ohm and megohms (millions of ohms).

If you're curious, AC resistance is called 'impedance', and is too complex to deal with here, as it is frequency dependent. Nor is it something you need to know about to do good wiring.

Third, DVOMs measure current – which is the amount of electricity passing a certain point, as opposed to voltage, which is the force with which the electricity is passing a point.

Think of the difference between a water hose with a nozzle that focuses the spray to a tight stream, and the same hose without the nozzle, with the water just gushing out. The hose without the nozzle may pass more water (current), but it will not have the force (voltage) of the tight stream from the nozzle.

The basic unit of measurement for current is the ampere, or amp for short. But an amp of current is actually quite a lot, so we often deal in amounts as small as thousandths of an amp – or milliamps. From this comes the name 'milliammeter', which is a meter for measuring very small amounts of current.

Now you know what all the letters in the acronym DVOM stand for. Kind of makes you feel glad all over, doesn't it?

Fancier DVOMs add more functions, signal frequency, temperature measurement, audible tones and lights. But all will offer the basic functions of measuring volts, ohms and milliamps.

What this means is that we can check our soldering with the ohms function, check AC and DC voltage and, if the need arises, we can check the current flowing in a circuit.

For our needs, I'm going to skip current measurement, as you can solder thousands of cables and never use that function. Google on 'using a DVOM' or 'using a DMM' if you want to learn more.

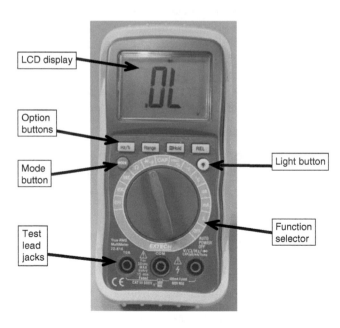

A typical DVOM is shown in Figure 2.2.1. Before going any further, I'd like to quote from the owner's manual for this DVOM: 'Improper use of this meter can cause damage, shock, injury or death. Read and understand this user manual before operating the meter.' Nor does the manual exaggerate; you really can kill yourself if you don't use a DVOM in a safe manner.

So my overview here is no substitute for reading the instructions that came with *your* DVOM. As the acronym says, 'RTFM' – which stands for 'Read The ******* Manual'!

Nor is my short explanation any substitute for common sense. Keep your hands away from electrical contacts and wear rubber-soled shoes. Have someone nearby with an insulated tool to knock you free of high voltage/current contacts if you're doing such measurement. Above all, think, plan and act carefully.

Figure 2.2.1   Parts of a DVOM.

Now that I've (hopefully) scared you into using it carefully, let's have a closer look at our typical DVOM. This unit costs about $100 and comes with lots of nice features I won't explain, because I want us to concentrate only on those features needed to wire cables and test them.

Our example is an ExTech model 22-816 that I got at Radio Shack when my old DVOM died. It's an RMS unit, which in a nutshell means that it measures voltage by averaging, and may have a built-in error value. Peak reading DVOMs are more costly and not needed for testing cables.

Don't feel that you're obliged to buy this, or any specific model of a DVOM, but you will need one of some kind to test your work. Shop around, and find a model that suits your needs and budget.

Since the LCD panel is exactly that, let's start by examining the controls below it. Then we'll go into using the three types of measurement useful for wiring: DC resistance, DC voltage and AC voltage.

On this model there are four option buttons (Figure 2.2.2). From left to right in the figure these are:

- Hz/duty cycle button – for wiring, we can ignore this button. With the DVOM in Hz/% mode, you can select Hz for frequency or % for 'duty cycle' – which can be loosely defined as the amount of time a signal is present vs. the total time being measured.
- Range button – this DVOM is auto-ranging, but you can select a specific measuring range by toggling this button, e.g. ohms or megohms. This is useful mainly for precision measurements.
- Hold button – this will 'freeze' a measurement on the LCD screen. Press once to freeze, once again to release.
- Relative (REL) button – this acts like the 'tare' function on a scale. When you press it, any residual voltage or test lead resistance is zeroed out, leaving the meter displaying all zeros, and ready to measure. This is analogous to the tare on a scale, which is set to zero out the weight of a bag holding the items being weighed.

Figure 2.2.2 Option/Mode/Light buttons.

The two other buttons in this section perform different functions. They are:

- Mode button – used in combination with the 'diode' function on the rotary function selector dial. 'Diode' is symbolized by a right-pointing triangle combined with a sideways cross (Christian style). In Figures 2.2.2 and 2.2.3, the function selector is pointing to the diode function. When diode is selected, toggling the Mode button switches between diode check and a DC resistance measurement that *beeps* when the resistance is low. Diode check is not useful for wiring, but a beeping ohms range is *very* useful, as you don't need to look at the DVOM to know that you have a good signal path.

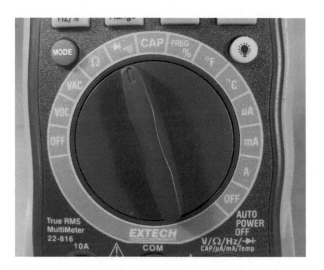

Figure 2.2.3   Function selector switch.

- Light button – turns on a backlight for the LCD display. On this model, you have to hold down the Light button for a second or two to activate the light. This feature helps to keep the light from being turned on by an accidental bump.

Next up, or down, if you realize that I'm describing the DVOM in descending order of the controls, is the rotary function selector dial.

One might say that this is the 'heart' of the DVOM, since it's by spinning the rotary function selector that we can adjust how and what the DVOM measures.

For our purposes, only the four functions on the left need attention. They are VDC, VAC, ohms and diode check/beeping ohms. Oh, and 'off' too, so we save the charge of the 9 V battery inside the DVOM. Make that five functions.

For all your work with a DVOM, the red test lead goes in the right-hand (red rimmed) jack and the black test lead goes in the center (black rimmed) jack (Figure 2.2.4). The only exception is for high amp tests, when the red lead moves to the leftmost jack.

Figure 2.2.4 Test lead jacks.

Since such measurements can be lethal if done incorrectly, I'd like to say that you should not try high amp measurements, and thus 100% of your work with a DVOM will be with the test leads in the position shown in Figure 2.2.5.

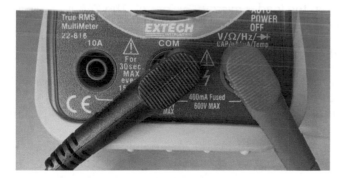

Figure 2.2.5 Test leads inserted.

Make sure your test leads are firmly inserted, and we're ready to work with the DVOM!

Our next series of shots describes the use of the four test functions above. I'm totally confident that you can use the 'off' function without any further instruction!

The first exercise is to measure the DC resistance of different types of wire, so I'll set the rotary function selector to ohms or Ω (Figure 2.2.6). I could also use the diode/beep function just to the right of the Ω), but not all DVOMs have it. I'm also going to call the rotary function selector an RFS, because I'm tired of typing it!

Figure 2.2.6   Set to ohms function.   **Figure 2.2.7**   First reading in Ω function.

So I cranked the RFS over to Ω and the cryptic message in Figure 2.2.7 appeared on the LCD panel. The O.L refers to 'open load' – which is the DVOM's way of saying, 'Dude, I have nothing to measure!'. Electro-tech speak often refers to a 'load' or object-under-test. Since there is no object-under-test yet, the load is missing, or 'open'. The DVOM is kind enough (and smart enough) to tell us this.

There are three more bits of info here that are easy to miss:

1. In the upper left corner of the LCD is 'auto', which shows that the meter is in auto-ranging mode.
2. Under the O.L is 'MΩ', which indicates that the DVOM is currently in the megohms range of DC resistance.
3. Between the O and L is the decimal point, or '.' – *very* easy to miss, as the decimal point will *change* its position according to what you're measuring. So keep a sharp lookout for that little dot!

These types of information will *always* be displayed at the same location on the LCD. Or, all but the decimal will, and it moves in a logical manner. Learn to make looking for them part of how you see/read the meter.

But I want to actually measure something! Let's see what happens to that O.L when I touch the test leads together (Figure 2.2.8).

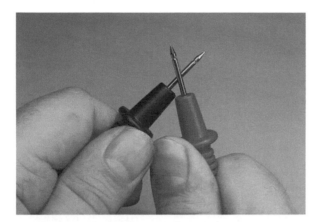

Figure 2.2.8   Touch test leads.

I'm touching the test leads firmly together; they are dry and clean. The LCD display flickered and quickly settled to the reading seen in Figure 2.2.9.

Almost everything's changed now on the LCD. We're still in auto, but now we're measuring in the Ω range, not MΩ. The O.L is gone, and we see 000.4 on the display. That means the test leads themselves have an internal DC resistance of four-tenths of an ohm – pretty low, but we want to be really accurate. How to adjust for the test leads? Oh, the REL button (Figure 2.2.10), that's right!

Figure 2.2.9   Resistance of leads.

Figure 2.2.10   Use of the REL button.

It's what you *don't* see here that's critical to understand – I'm *still* holding the test leads firmly together with my left hand! If I didn't do that, I'd go back to that boring O.L. So while holding the aforementioned leads together, I pressed the REL (relative) button and, lo, and behold all zeros.

Now the test leads and the DVOM are a calibrated unit, ready to measure accurately. Let's go find some wire!

A good subject for our first test is a short length of two-conductor shielded audio wire (Figure 2.2.11). It's typical of internal studio wire.

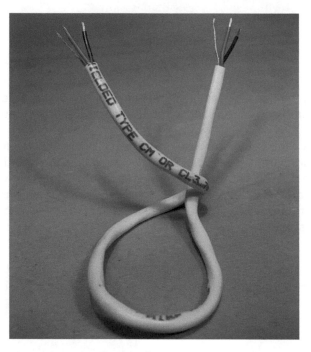

Figure 2.2.11   Short audio wire.

With the wire resting so I can press down on the strands with the test leads (Figure 2.2.12), I get the reading shown in Figure 2.2.13.

Figure 2.2.13   Short wire resistance.

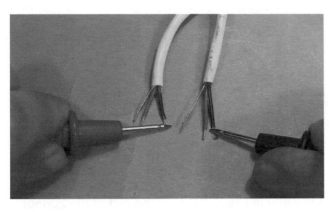

Figure 2.2.12   Test short wire.

We can see there's not much DC resistance – only 0.3 Ω. This makes sense, because it's such a short piece of wire.

Let's try some different kinds and lengths of wire, to see if there are changes in what we measure.

In Figure 2.2.14 I'm measuring an equally short length of coaxial video cable. It's the kind of wire that would go to a TV cable box. I get the result shown in Figure 2.2.15.

Figure 2.2.15   Video wire resistance.

Figure 2.2.14   Video wire test.

The coax cable has a higher resistance – 0.8 Ω. That's still pretty low. Let's go measure something longer!

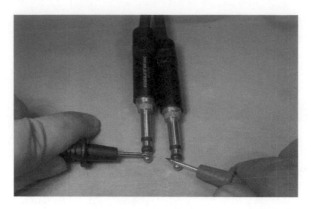

Figure 2.2.17 Twenty-foot cable resistance.

Figure 2.2.16 Test 20-foot cable.

Hiding out of the shot in Figure 2.2.16 is the rest of the 20-foot guitar cord that I'm testing. The result is seen in Figure 2.2.17. Our invisible 20-foot guitar cord measures out to 1.5 Ω – a nice, healthy reading.

In general, your resistance readings should be 2 Ω or less, unless you're measuring a very long wire run. Higher readings may indicate cold solder joints, defective wire or tarnished connectors.

But this has all been much too safe and tame. Let's measure something a little more lethal!

Just to be sure you remember those acronyms – I'll say it the long and boring way: 'Set rotary function selector to volts AC' (Figure 2.2.18). We're going to measure some plain vanilla 120 VAC from my wall outlet.

Figure 2.2.18 Set the RFS to VAC.

OK, I lied – we're measuring the same 120 VAC (Figure 2.2.19), but I found it a lot easier to bring an extension cord and a couple of female three-way AC splitters up to the camera table than it would be to show where the whole mess plugs into the wall outlet. Same voltage, same current – it'll still kill you if you mess up.

Notice that the test leads are fully inserted, and my hands are well away from any possible contact points.

The actual reading of my nominal 120 VAC is shown in Figure 2.2.20. It's a tad high, at 121.9, but well within spec. Typical US house AC readings will vary from as low as 105 VAC up to 130 VAC. Electrical power supply companies are not known for delivering ultra precise power – but for us in the US, it's there all the time, like the air we breathe. Could you imagine the quality of life in a country where that is not a given?

So now you have some idea of measuring resistance to check your wiring. And you also (I hope) know how to measure 120 VAC without hurting yourself.

You can check to see if there's electrical power for lights and soldering irons, maybe even some fans to blow the rosin smoke away.

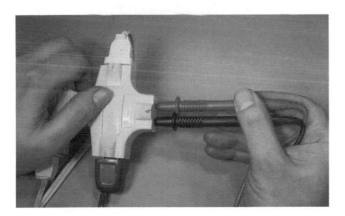

Figure 2.2.19   Measure 120 VAC.

Figure 2.2.20   The 120 VAC reading.

Figure 2.2.21   Set the RFS to VDC.

There's one more test I want to show you – DC voltage. To do this, we have to invoke the RFS again. That's 'Set rotary function selector to volts DC' (Figure 2.2.21). I just had to say that to make sure.

Now we need some DC electricity – what can we use for a source? How about an AA size flashlight battery, a great source for VDC (Figure 2.2.22). Just open up your Maglite and flip one out.

Figure 2.2.22   A 1.5 VDC source.

In Figure 2.2.23 I'm testing the AA battery, but wait – I've made a mistake and reversed the test leads. Normally the red lead would go to the positive terminal. What will happen because of this? Look at the reading in Figure 2.2.24. Aha! See the minus sign? Also the 'Auto', 'DC' and 'V'? So my measurement is accurate, but inverted because I flipped the test leads – another thing to look out for.

Figure 2.2.23   Test AA battery.

Figure 2.2.24   AA battery voltage.

There's also a caveat that since a DVOM does not put a real-world 'load' on the battery, as a light bulb would, the DVOM will give a falsely high reading.

Real battery meters have built-in loads. Test a bunch of dead/semi dead batteries and you'll soon get an idea of what your particular DVOM can show you. At the very least you can sort out the totally dead batteries with a DVOM.

This concludes my incomplete introduction to the world of DVOMs, how to use them, and why to use them. There's a lot more to say about them, but that info is already out there – you just have to go look for it.

If I've given you a taster and made you want to learn more, I've done my job. As I said at the beginning, Google on 'using a DVOM' or 'using a DMM' if you want to learn more. 'The truth is out there', but I don't have room here for all of it – just enough to whet your appetite.

Balanced and unbalanced audio and AC power

After much skull scratching and soul searching, I decided to combine several concepts into one section, because they are so intimately interconnected. No, not *that* intimately, they're just good friends.

So in this section I'll talk about unbalanced and balanced audio, unbalanced and balanced AC power, and the best ways to wire and clean up the sound (and picture) of your studio/disco/home theater/whatever.

Unbalanced/balanced audio

Let's start with audio; a nice, simple bit of audio – a sine wave. Some of you may have seen a sine wave on an oscilloscope or in a picture. They all look more or less like the one in Figure 2.3.1.

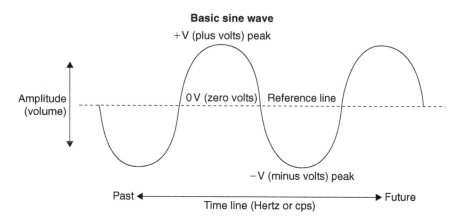

Figure 2.3.1 Basic sine wave.

Since the sine wave is AC (alternating current), it will start at 0 V (zero volts), rise to a positive peak, then reverse itself, cross the 0 V reference line again, and rise (inversely) to its negative peak. Or it will do what I've shown here: start negative and flip positive. And it will keep doing this, over and over, until we get bored and turn it off.

If the sine wave repeated this action 1000 times in a second, we'd say it has a frequency of 1000 Hz (hertz) or, in older terminology, 1000 cps (cycles per second). We audio folks got tired of saying 'see-pee-ess' and renamed the unit of measurement 'hertz' as it's shorter.

Higher frequency sine waves will appear more squished together horizontally; lower frequency sines will look more spread out. The reason for this is that the horizontal axis in an oscilloscope is the 'time base' – it shows the progression of the waveform from the past into the future. The more times a signal reverses polarity, the higher its frequency and the more reversals present in a given period of time.

All sound (almost) is made up of complex combinations of AC (alternating current) waveforms, most of which are not sine waves. The only exception is a DC (direct current) pulse, which will make a one-time 'click' when connected to a speaker or headphone, but not much else.

We use sine waves for measurement because they're easy to quantify.

I hope you are now fine with sine, as it were, and ready to see how this applies to real-world situations.

There are only two ways that an analog audio signal can be carried along in a wired connection. For the sake of brevity (and sanity – mine), I'm not going to expound on digital or RF transmission of audio.

The simplest way for an audio signal to be carried on a wire is as an unbalanced signal. This means that there is a center conductor (hot), and (typically) shield and ground are combined in the outer layer of the wire. So half of the signal path is (sort of) shielded by the outer layer, and the outer layer itself is tragically vulnerable to interference from sources in the outside world.

What this means is that unbalanced audio is basically limited to runs of 20 feet or less, and even then it lacks the ability to null out induced noise, hum and the other crud we encounter with great ah, frequency.

Balanced audio, on the other hand, can survive runs of hundreds of feet, so all pro audio facilities use balanced mic lines, balanced transmission lines, and do most of their internal wiring in a balanced manner.

We'll explore what unbalanced/balanced wiring is after we take a quick peek at a couple of guitar plugs to show you the physical difference between balanced/unbalanced connectors.

I'm going to recycle some pictures here from Section 3.1. And being lazy, I'm also going to recycle some of the text, as I worked very hard to make the differences clear, and cannot find any way to make things more lucid than what I've already written.

What you'll see next are the solder tab ends of a stereo and mono guitar plug, followed by the 'business ends of the plugs that actually get inserted into guitars, amplifiers and other gear. And if you guessed that the mono plug is unbalanced, while the stereo plug can be wired balanced, you get a gold star!

There are always caveats and this example is no exception. The so-called 'stereo' guitar plug can be wired as a single balanced connection, or *two* unbalanced mono connections that share a common ground. So don't assume, always check.

A close-up of the two solder tabs on a stereo male guitar plug is shown in Figure 2.3.2. I've drawn two arrows to show exactly what part(s) I'm talking about. The longer part, that extends to the upper left in this picture, is both a strain relief for the wire and the part that the shield/drain gets soldered to.

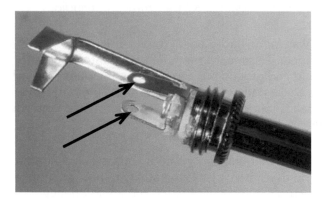

Figure 2.3.2   Solder tabs of stereo male guitar plug.

Let's call the two tabs I show the 'upper' and 'lower' tabs in this picture. The lower tab goes down to the tip of the plug. It's the high/hot conductor.

The upper tab goes to the ring of a stereo plug, but is *omitted* (not present) in a mono plug. It's the low/cold conductor. As a general rule, tip is high, ring is low, and the long barrel of the plug is used for drain/shield.

Since I want everyone to be totally clear on the difference between stereo and mono plugs, I've got a couple of side-by-side comparisons ready.

These pesky plugs are so shiny I had to put some white artist's tape behind the solder tabs, so you could see them against the strain relief behind them (Figure 2.3.3). I hope it's all clear. On the left is a mono plug with one tab. On the right, a splendid example of a stereo plug with two solder tabs.

Now that we're straight on the tabs, let's see the whole plug (Figure 2.3.4). Here we can see the business ends of our plugs – mono on the bottom and stereo on the top. Notice the ring on the stereo plug? That's the part the low conductor is connected to – and is clearly omitted in the mono plug below it. So one tab = no ring, mono plug. Two tabs = has ring, stereo plug. And remember, a 'stereo' plug can be wired as unbalanced stereo or balanced mono – the wiring will look the same.

Figure 2.3.3   Mono/stereo comparison – 1.

Figure 2.3.4   Mono/stereo comparison – 2.

Figure 2.3.5 illustrates how an unbalanced mono plug is connected to a one-conductor shielded wire. Or, in other words, to a wire that has one internal conductor which is surrounded by a shield, that also functions as the 'low' side for the audio signal.

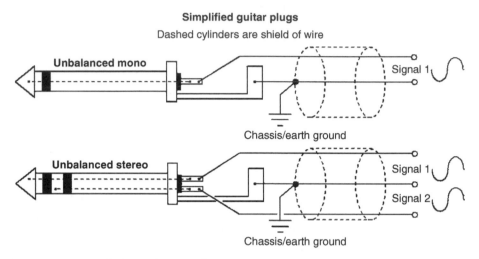

Figure 2.3.5   Unbalanced guitar plugs.

The same figure also shows how an unbalanced stereo plug is connected to a two-conductor shielded wire. That is, to a wire with two internal conductors, both of which are surrounded by a shield. The two signals are discrete, but share a common ground.

The two signals could be left and right of a stereo signal, or they could be two totally unrelated signals, so the nomenclature of 'stereo guitar plug' is ubiquitous, but not really accurate.

Hopefully, with the aid of the previous picture, you now have a clear concept of how unbalanced audio is connected. The same rules apply whether you are using guitar plugs, RCA plugs or whatever the 'plug de jour' happens to be today.

But what about balanced audio? Why is it called 'balanced', and how does it differ from an unbalanced signal? This is where we come to some very clever voodoo.

Balanced audio is created by splitting the audio signal into two separate but equal parts, and then inverting (flipping) the phase of one of the two.

Your instantaneous question may be 'Why bother?'. The reason is that when the in-phase and the out-of-phase signal are properly recombined (by uninverting the flipped phase side in a particular way), the result is that our desired audio signal is not only amplified, but any stray noise it has picked up is immediately nullified, leaving only the pure signal.

This is such an important concept that I'm going to repeat it in different words, hoping that it will embed itself deeply in your minds.

Balanced audio reduces or eliminates unwanted noise picked up in wires by flipping (inverting) the phase of one of the two conductors that carry the signal. When the signal is properly recombined, its amplitude (volume) is increased and the unwanted noise is nulled out.

Yet another way to describe this is that when the plus (+) noise is summed (added) to the minus (−) noise, the result is *no* noise. Or at least very little noise.

What this means is that balanced audio runs can be hundreds of feet long without degrading the signal by adding noise to it. Pretty cool, huh?

Figure 2.3.6 shows a balanced mono guitar plug, and also the noise-cancelling concepts we've talked about above. Pay particular attention to it, as the subsequent discussion in this section is based on you having a clear understanding of how balanced audio works.

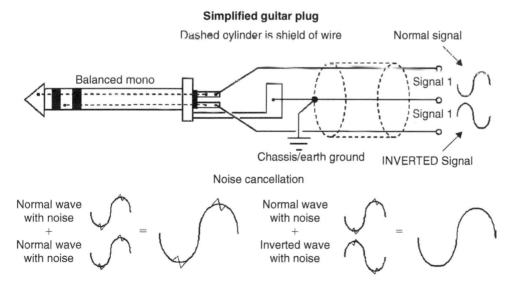

Figure 2.3.6   Balanced connection/noise cancellation.

Hopefully, I've now tossed this information at you in enough different ways that you've got a decent grasp of it. Let's put it in still another way.

Balanced audio lines help cancel out interference of many types. Not only hum (ground loops), but also buzz (60 Hz harmonics), thermal sound (white noise), digital clock jitter and lots of other bad stuff, too numerous to mention.

Next up is an example of a typical balanced +4 dBu audio connection, the kind of connection you might make from a pro-level recording console to a pro-level audio recorder – analog or digital. This example is shown in Figure 2.3.7 for an XLR type (three-pin) connection.

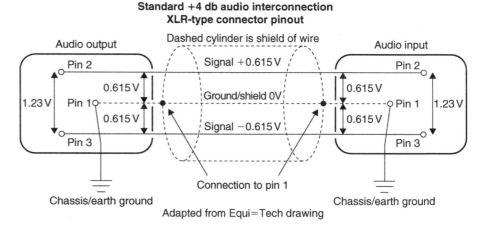

Figure 2.3.7 Balanced +4 dBu XLR-type connection.

You don't have to pay too much attention to the voltage values – they represent an ideal you might see on your DVOM, on a clear day with a favoring tailwind.

The only function of the voltages in this diagram is to give you some idea of what you might encounter in the real world, and reinforce the concepts of balanced audio.

With luck – and attention on your part – you've now seen the advantages of balanced audio. You will restrict your unbalanced connections to short runs and, if given the option, always wire gear in balanced mode, right?

Now we come to the real mind-blowing part. *Electrical power is basically an audio frequency signal!* We're all painfully familiar with the sound of 60 Hz hum – it's ubiquitous. No matter where you go, you hear it – anywhere within the AC power grid, and often up to several miles away from it.

But is our regular run-of-the-mill 120 V, 60 Hz electrical power distributed in a manner similar to balanced audio in a studio, to reduce noise pick-up? *No!* Regrettably, all standard 120 V power distribution systems are wired in an unbalanced mode – this makes them highly susceptible to picking up all kinds of crud!

Every time you hear 60 Hz (or any other noise) in an audio system, it's degrading the sound quality and robbing your amplifiers of power.

This brings us to our next section in this module. The truly observant among you noticed that the last figure included a credit to something called Equi=Tech. In the next part, you'll find out why that mysterious credit is there. Can you wait that long?

Balanced AC (electrical) power

Just as unbalanced audio lines are degraded by noise pick-up, so too are our AC power lines. We take AC power for granted, because the lights and appliances work. But computers, synthesizers, sequencers and all our other sophisticated electro-toys require much cleaner and higher quality AC power than your grandma's toaster.

There's an old computer programmer's acronym – 'GIGO' – stands for (G)arbage (I)n (G)arbage (O)ut. The quality and functionality of a computer program can be no better than the coding that created it. Bad code makes for a bad program. Or, in our case, bad quality AC power creates bad quality audio and video, jittery digital gear and ditto for all the other electro-toys.

Each electronic component does have some power filtering and smoothing ability built in, but it's often as minimal as the manufacturer can get away with. It's ironic in a sense. We spend tons of money on noise reduction, noise gates, dynamic noise filters and more, but we typically ignore the source of much of the noise itself – the AC power being fed to the equipment!

That 60 Hz note we hear is actually a real, musical note – a flatted B1, if you call 'middle C' C4. And the fact that our power grid is truly universally contaminated has the whole world singing the blues. That's just my minus 41 cents worth (the amount that B1 is flatted).

If balanced transmission is good for audio, why not use balanced transmission for the audio frequency signal that *is* our AC power?

Why not start at the source? Feed clean, balanced AC power to our gear? It will pay us back with higher quality performance, greater stability and its eternal gratitude.

This opens up a host of questions that I see swirling in your mind. Let's make a list of them and I'll answer quickly, so we can go on.

Q: Does this mean that the entire AC power grid of the whole world is mis-wired?! That's mind-boggling!

A: Yes, it means exactly that, from the viewpoint of providing optimal AC power. The only optimal AC power distribution is in private facilities that have created their own supply.

It also means that we are faced with the prospect of (eventually) rewiring the whole world if we choose to do so. However, we've already done this task twice. You quickly ask me, 'When was that?'

Once, when we converted from DC power to AC power, and again when we converted from two-pin AC to three-pin AC. How quickly we forget the techno struggles of yesteryear.

It is likely that balanced AC power will remain a private sector decision, as I know of no town, city, state or country contemplating its deployment, although there are excellent reasons why they should.

Q: OK, you've convinced me. How do I get some balanced AC power for my studio?

A: There are several companies who sell balanced AC power supplies in various sizes. You could even make your own, if you were determined, a good electrician, and wanted to reinvent the wheel.

For the sake of brevity, and my carpal tunnel syndrome, *balanced AC power will henceforth be abbreviated as BACP*. If you don't grasp that, give this book to a friend!

Q: Don't worry, I understand! I want that BACP right now! Isn't it expensive?

A: No, not really. Musician-size units start at around US $1000 and scale up in cost according to the size you need. BACP can even *save* you money! I'll explain this later.

Q: Is it hard to use?

A: No, you plug the BACP supply into your regular AC outlet. Then you plug your gear into the BACP. Simple.

Q: Can I use my UPS (uninterruptible power supply) along with the BACP supply?

A: Yes, but you *must* plug the BACP supply *into* your UPS supply. So you better have a honking big UPS – none of that wimpy stuff. The hook-up goes like this:

(AC power at wall socket) → (UPS) → (BACP) → (Gear). Clear now?

Q: There are a lot of 'power conditioners' on the market. Do they all create BACP?

A: Absolutely not! Only a handful of companies make real BACP supplies. Other 'power conditioners' do, in fact, clean up some of the gunk on your raw, nasty, polluted AC, but only those units that give a power output of (normal 60 VAC) → (0 V neutral) ← (inverted 60 VAC) are true BACP units.

Only those units provide the noise cancellation of BACP. I'll explain this in detail shortly. My goal in this Q and A section is to give you an ultra-fast overview and then fill in the details later.

Q: You said that BACP would cost me at least US $1000. Then you said it could *save* me money?! Explain!

A: Ah, very good! You really *are* paying attention. By using BACP most of the customized wiring for a studio is eliminated. You can use stock cables and connect ground everywhere.

You don't have to create a star-ground system. In fact, the more places you connect ground, the quieter the system becomes, as more noise is nulled out.

Since custom star-ground systems are, by definition, time intensive to wire, they are also costly to create. So the cost of your BACP may *save* you several times your investment in it, by reducing your wiring costs.

In other words, your $1500 BACP will eliminate a significant part of your (potentially typical) $8500 (or more) wiring costs. Not all, of course. You still have to wire your gear together.

But now you can use stock cables, mix and match stock with custom work, and only custom wire what is needed. Cheaper, faster and it sounds better too.

Q: What if I can only afford BACP for some of my gear?

A: Excellent question! Take a look at http://www.equitech.com/faq/somequip.html. But you have to promise me not to peek at it until you've read the rest of this section. Deal?

Introducing Equi=Tech

BACP is not a new idea; it's been around for decades as an esoteric laboratory concept for diehard audiophiles. Back in the early 1990s, Martin Glasband created the first commercially marketed BACP supplies through his company, Equi=Tech. The company URL is www.equitech.com. There is a *vast* amount of useful info on this site! Much more than I can include here.

Nowadays other companies are producing BACP supplies. There is some question about the legality and ethics involved, since none of them pay (to my knowledge) any royalty to E=T, which holds the patents for developing BACP. This was confirmed by E=T itself when I queried them.

As an audio engineer, and tech, I've followed the testing and growth of BACP for over 25 years. In all that time, I've never met Mr Glasband, I have no stock in Equi=Tech, and don't make a penny off the sale of E=T equipment. But I've talked to many E=T BACP users and a few facts stand out:

- No user I've ever met has given up their BACP supplies once installed.
- All the users I've met said they experienced a dramatic reduction in noise, and a huge increase in sound and video purity.
- The late Dave Smith (head of R&D for Sony Music Studios at the time) once said to me that BACP made such a difference in how effects units sounded that he almost wanted a huge power switch, so he could flip back and forth, and choose the sound he liked best for what he was processing at the time.

- The client roster at E=T reads like a who's who of the audiophile and pro audio industry. Check out this link: http://www.equitech.com/ourclients/someclients.html. This roster represents millions of dollars quietly invested by the cognoscenti to improve their sound – almost as a 'secret weapon'.
- Any company that makes gear used by NASA, Digidesign, the FAA and the Cirque du Soleil has to be doing something right. It can't *all* be smoke and mirrors.

When I started writing this section, I wanted to quote some of Mr Glasband's articles and other works on the E=T website. Mr Glasband kindly consented, asking only proper credit when due. I'll be using snippets of text and some illustrations; in these illustrations, the credit 'adapted from Equi=Tech drawing' means that I redrew a diagram from the E=T website for greater clarity. All other illustrations are entirely my own work.

BACP theory and practice

Now that I've given you a general outline of what BACP is, and why to use it, we can get down to the nitty-gritty details.

The 120 VAC single-phase AC coming out of your wall socket is often a split-off from 220 VAC two-phase or 220 VAC three-phase AC that is installed as the primary power source to a facility.

I'm not going to describe the 220 VAC interface, as you will not be using it, unless you're rewiring a major installation for many kilovolts of power. My point is, by the time you get it, that 120 VAC is always unbalanced, no matter what the configuration of your primary building power is.

The 120 VAC power circuits typically available are illustrated schematically in Figure 2.3.8. The funny looking thing on the left side of the diagram is the electrical symbol for a transformer. The rounded squiggles represent the coils of conductors in it and the two vertical lines are the (typically) iron core they are wound around.

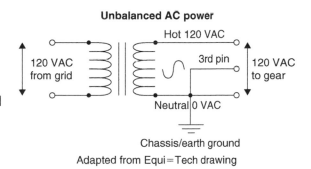

Unbalanced AC power

Adapted from Equi=Tech drawing

Figure 2.3.8 Typical unbalanced AC.

Whether you realize it or not, the other end of the wiring to your local 120 VAC wall socket is the output of some super-sized power transformer upstream from you in the power flow. It could be a local splitter transformer from your utility company or some other configuration. But it's always there – it has to be there.

On the left side, our transformer is being fed AC power from the grid. On the right side is what you get – the AC power running your gear, lights and popcorn maker.

Notice that although there are three conductors, ground and 0 V are tied together. This is a safety precaution, so that if there is a short to chassis of the 120 VAC, it will blow a fuse or trip a breaker, instead of making you light up, smoke and smell bad.

So although there are three conductors, there are really only two signal paths – 120 VAC high and 0 V low/ground. Unbalanced. Bad.

How can we fix this? Martin Glasband found (one might say rediscovered) a simple and elegant solution. If we reference ground to a 'center tap' on the transformer, we can still get the 120 VAC we need, but now the 120 VAC is fed out in a balanced manner, well able to overcome myriad forms of contamination. It is illustrated in Figure 2.3.9.

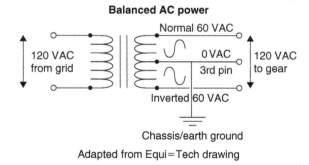

Figure 2.3.9 Balanced AC.

Since the center tap of the transformer is tied to ground, we still have protection from short circuits, but now each leg of the transformer is carrying 60 VAC referenced to ground.

It's +60V on the former high leg and −60V (inverted) on the former neutral leg, but the voltage potential across the two legs is still 120 VAC of balanced AC power. What's good for audio is good for AC power too. Balanced. Happy, clean, quiet, safe. Nice.

What this means is that any BACP supply is, itself, built around a special (center-tapped) transformer.

We plug the BACP transformer into a source that feeds it unbalanced AC power, from the utility company's power distribution transformer. The BACP supply then converts that dirty, old, unbalanced AC into nice, clean, balanced AC for our enjoyment.

We're using a (special/center-tapped) transformer and related circuitry to correct a problem inherent in the (upstream) transformer feeding AC power to it. Clear?

Here are some quotes from Martin Glasband:

'Ideally, the correct type of AC power for audio gear in the United States is 120-volt, two-phase (balanced) power. Commercial use of two-phase (equi-potential) AC isn't new. In fact, it is probably the first type of AC power ever put into widespread use in this country (Chicago in the early 1900s). Nevertheless, its commercial use is all but dead these days.'

'Properly grounded 120-volt two-phase wiring systems aren't mentioned at all in the National Electrical Code (NEC) [thanks to E=T and MG this is no longer true]. However, this unfortunate fact didn't deter a few innovative engineers and studio execs from employing a 120-volt equi-potential AC system at the Zoo Studios, Studio City, CA.'

'Soon after its opening, a studio musician plugged in his Marshall amp and declared it was broken because there was no hum. Much to his surprise, his amp worked just fine. He mentioned to the studio engineer that this was the first studio he'd ever worked in where his amp simply made no noise at all. It should also be noted that it was not necessary to drive a single ground rod. That's pretty clean AC.'

(http://www.equitech.com/articles/rep1.html)

The following are quotes from Roger Nichols, a well-known (and respected) audio authority, taken from the E=T website. He's also a talented (and very funny) writer – worth looking up.

'Spike, noise, surge, swell, transients, harmonics and sag are not the names of members of a new alternative rock group; they are characteristic problems encountered in power management.'

'AC power is often the most overlooked area in recording studio design. If you were a farmer and your horse was your livelihood, you would probably pay attention to how well he is doing.'

'AC power is the main source of your income, and also the primary cause of all the hums and buzzes you must deal with on a day-to-day basis. They say that if you build a better mousetrap that they will beat a path to your door. Well, just wait until you have the quietest studio in town and see how fast everyone wants to work there.'

'Quiet grounding schemes in studios sometimes border on the occult. I asked one studio why they had a water cooler in the control room with no water in it. The said that, for some reason, when the water cooler was plugged into the same branch circuit as the guitar amps, there was less hum in the amps. I unplugged it once. They were right.'

'Grounding circuits were never meant to carry current except during a short circuit. Objectionable ground currents are those that will provide you with a shock. Anything less than that is OK as far as Underwriters Laboratories is concerned.'

'We have all experienced ground loops in the studio. The really bad ones, with hum levels above the signal level, we try to cure. The ever present little hums, that make the DAT meters stick one segment up from the bottom, we try to ignore.'

'We try breaking grounds in balanced cables at one end so that we do not have multiple ground paths for ground loops. We lift chassis grounds with special plugs and make sure that metal chasses do not touch each other. *If we removed the currents from the ground, then we would have no current to loop* [my emphasis – JH].'

'With balanced power, you can use any type of grounding configuration you wish. Star, schmar. You can leave the grounds connected at both ends of your audio cables. You can throw away all of the ground lift adapters. You can finally plug everything in the way it was meant to be plugged in.'

(http://www.equitech.com/articles/power.html)

Whew, sounds pretty compelling to me, what do you think? Roger Nichols is a bit over the top at the end – if you go BACP, you need to *put back* all those grounds you lifted when trying to star ground. The BACP scheme works best when ground is connected at all points.

I could go into a lot more detail, but all the info is on the web – at the E=T site, or easily available through Googling. Try searches like 'balanced AC power' or 'star grounding'.

Speaking of star grounding, I keep referring to it, but I've never said what it is or how to do it – I'd better fix that right now.

Star grounding

Prior to the development of BACP, the only effective way to combat ground loops was to create a star-ground system. The name comes from the concept that *all* gear is grounded at *one* (and *only* one) central point. The ground conductors radiate out from that central point like the rays of a star – hence the name.

Every other path to ground should (ideally) be ruthlessly eliminated. This includes chassis-to-chassis contact, contact through metal rack rails, and contact through third-pin AC plugs.

Each unit would (typically) have its audio ground lifted at input and carried at output. The third pin on the AC plug would get a ground lifter, and a ground wire would be run from a (tested) ground point on the unit to the central point of the star ground.

Sound complicated? Time intensive? Costly? You betcha! But it worked. However, I have to add 'illegal' to that list, as lifting the ground on the third pin is against electrical code.

And I also have to add 'potentially lethal' because if the third pin is lifted, the star-ground conductor is detached and the unit has a 120V short to chassis – well, you could die. Admittedly, the chance of all three happening is remote – but why gamble?

I've wired pro-level star-grounded studios and gone through all of the above. Sometimes results were spectacularly good – but at a high cost in both labor and money.

If I had to do it again, I'd get a high-caliber pistol and hold it to the studio owner's head, until s/he agreed to do a BACP installation.

Your choices

In wiring your own facility, you have basically three choices. Of these, I suspect the economics involved will limit you to two. Here they are:

1. Connect ground everywhere *without* using BACP. This is usually what happens anyway. It gets you up and running, but things are noisy, gear acts flaky and stupid, ground loops come and go. Sounds like your typical home studio, right?
2. Create a rigorous star-ground system. It takes a lot of time and labor, but will give you cleaner sound, fewer loops, and more reliable operation. It's difficult to maintain system integrity, as the star-ground scheme is typically broken when gear is added or changed.

I started this book in 1991, and at that time I was a star-ground fanatic! I sank ground rods, drilled I-beams – the whole nine yards. There was no better technology and I pushed star grounding to the max.

I had special non-conductive rack rails made out of wood, and made space between units mounted in racks to avoid chassis-to-chassis contact – and all of it helped, often radically. But these days, I think BACP is the better option – which brings us to the final point:

3. Connect ground everywhere *using* BACP, If I was a studio owner, with a limited budget, I'd connect ground everywhere, get running and save my pennies for a BACP supply.

When I can afford it, I plug in the BACP, and I don't have to rewire anything! That's a lot better than wiring a star-ground system and then *rewiring* it for BACP.

If individual units give me a hard time, before I go to BACP, I try lifting ground(s) on just those units and keep track of what I've changed, as it will have to revert to the original state when the BACP goes in. Or I use audio isolation transformers to lift the ground for me. Do a Google on 'audio isolation transformer' – get *big* ones that can deal with high-level peaks and not saturate.

So here's my twenty-first century recommendation for twenty-first century power needs and twenty-first century audio/video production. Wire ground everywhere. Install BACP as soon as possible – ideally when building your facility. If you have to wait for BACP, deal with problematic gear on an individual basis.

There's so much more I could say – but I have to save space for the connector modules themselves. And spreadsheets – we have to talk about them as well. They're so cool when you use them right.

2.4

Using spreadsheets

Back in Section 1, I talked about organizing your information. One of the best and most compact ways to do this is with spreadsheet software.

For those of you who are already familiar with spreadsheets this will be a bit elementary, but I can't assume that all my readers know how to use spreadsheets – or even know what they are.

Figure 2.4.1 shows what a typical spreadsheet looks like before you start using it. It is laid out lengthways to fit the page better.

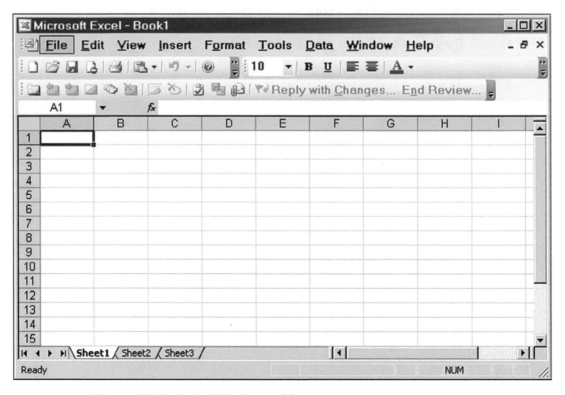

Figure 2.4.1 Typical blank spreadsheet.

All spreadsheets pretty much look like this before you start working with them. There are lettered vertical divisions called 'columns' and numbered horizontal divisions called 'rows'. The blank rectangular areas that are created by the intersection of the division lines are called 'cells'.

To make life even more potentially complicated, an Excel file can have several discrete spreadsheets open at the same time in different 'tabs'. In Excel, these tabs are called 'sheets' and if you fill up the first three, you can just add more.

You can have a staggering number of cells in a spreadsheet – in this version of Excel (2003), up to 230 columns and more than 50 000 rows. Your mileage may vary with other versions, or other spreadsheet programs.

For wiring purposes, we'll use a tiny fraction of these, so don't be intimidated. We also don't need to use any of the fancy features of spreadsheets – but don't get me wrong, you can if you know them.

However, for wiring, the main purpose of the spreadsheet is to store information in an organized manner. You don't need tables, or formulas, or embedded graphics. In fact, the main semi-advanced functions you'll use are the 'sort' option, the 'hide/show' option, the 'freeze panes' option, and the 'sum' option. More on these functions later.

There is absolutely *no* reason to work exclusively with Microsoft Excel. Our work can be done in any modern spreadsheet. I just used it as an example because it's so common – like weeds.

Open Office has a good spreadsheet that's free; another excellent free spreadsheet is Gnumeric, which is smaller and faster. Both work on Macintosh and Linux, as well as Windows. Nothing is perfect – there are some bugs, but no showstoppers that I know of.

You can find this cool, free software at the following links:

- Open Office = www.openoffice.org/
- Gnumeric = www.gnome.org/projects/gnumeric/downloads.shtml

If you think of each cell as a cubbyhole for information, that's a good start. The cell boundaries keep your info discrete, so it can be very tightly packed. It can also be copied, pasted, modified, deleted, duplicated and sorted, without fear of messing it up – if you're careful.

This is the main advantage of spreadsheets for our use – they can be updated and modified quickly, as opposed to manually erasing and then rewriting what you put down by hand. A written list is a great way to feed info *into* a spreadsheet; once you've made the effort to do that, the spreadsheet will be light years faster to work with than paper lists!

One very important concept is that the size and attributes of cells can be changed at any time, either for a group of cells or the whole spreadsheet. Bigger/smaller, wider/narrower, taller/shorter, color, font size and style, and many other attributes are all silly putty in your hands. Makes you feel almost God-like, doesn't it?

Let's look at a very simple spreadsheet, and then move on to more complex ones.

The first bit of spreadsheet that I present for your edification is one I made for a company that provides fancy TV cable service to upper-crust motels. Call the company XYZ TV to avoid complications. I was acting as a field tech for XYZ TV, and they had no organized way to report which rooms in what hotel had problems, what the problems were, and what the resolution was for each problem. Being lazy, I wanted to keep all the info about a given hotel in one place, so I made the spreadsheet shown in Figure 2.4.2.

The left side of the sheet has the type of info I want to find out and the center/right part is a place to write the information. In my example there are no entries for the answers, as I wanted to show the spreadsheet as if it were ready for use. I designed this particular sheet to print vertically on 8.5 inch by 11 inch paper, because it was so simple. Complex spreadsheets are typically printed horizontally (landscape mode) on 8.5 inch by 14 inch legal sized paper.

This spreadsheet is designed to be filled out either by hand or on a computer. Since the answer to each question has been given its own cell, data can be manipulated easily once computerized. And the printed version is perfect for manually gathering/writing the data in the same way you'd see it in the electronic version.

All of the spreadsheets in this chapter are available for download from the Focal Press website; just go to http://books.elsevier.com/companions/9780240520063, download the spreadsheets and print them on legal sized paper for the full impact of each layout. All but this one that is – this is letter sized.

Now that we've talked a bit about the functions of spreadsheets, and I've shown you this basic example, hopefully things are starting to make more sense, and you're getting a little more comfortable with the concept of actually creating your own spreadsheet.

Hotel Name	
Street Address	
City	
State/Zip	
Phone Number	
Contact Person 1	
Contact Person 2	
Additional Info	
Down Rooms	
Room/Issue	
Resolution	
Room/Issue	
Resolution	
Room/Issue	
Resolution	
Room/Issue	
Resolution	
Room/Issue	
Resolution	
Room/Issue	
Resolution	
Room/Issue	
Resolution	
Tech Name	
Date	

Figure 2.4.2   XYZ TV data collection sheet.

The next example is a spreadsheet I made to wire a guitar effects rack for my friend Sam (Figure 2.4.3). He has two medium-sized roadcases of guitar effects: a Mackie mixer and a fancy guitar switcher, called a Switchblade.

The switcher allows him to pre-configure different chains of effects, without having to re-plug things each time. Push a button and all is instantly re-routed. As you can imagine, this saves him immense amounts of time on stage and allows him to perform much more gracefully.

Sam's three racks are connected by big, honking multi-pin connectors, called DLs. The DL is a sophisticated connector; the model we used has up to 96 contact points. If we ever make it to *AWG Volume 2 – Advanced Connectors*, you'll be sure to meet the DL.

The goal I achieved for Sam is that all connections between his racks are on the multi-pin DL plugs. When it's time to set up, he plugs in the DLs, goes click, click, click to lock them, and only has to plug in his guitar, his foot pedal and AC power to be ready to play. A set-up that would take hours to connect is ready for action in a few minutes!

Further, each figure will overlap a bit if needed, to help you keep your place in all the columns I had to create.

At first glance, it reads like I'm talking techno-Martian here, right? But remember, this is an attempt to condense a lot of info into a tiny space, so abbreviations abound. And since there is no standard abbreviation code for wiring, I had to make up my own as I went along – just as you will, although you are welcome to borrow my terms.

In Figure 2.4.3 let's work our way across the top few rows, from left to right, and see if that clears things up a bit:

- WIRE RUN LIST 081301 – the run list as of 13 August 2001.
- DESCRIPTION – the name/function of each wire. As an example, PCM-42 DLY IN is PCM-42 delay line in. It's a mono unit, so there's only one in. REV 1 means that this is my first revision of the list.
- RACK 1 > GEAR – my shorthand for the connections that are internal within Rack 1, going to/from the Switchblade and to/from the effects gear mounted in Rack 1. Since the Switchblade is itself mounted in Rack 1, none of these wires have to go out to a DL for interconnection to another rack. All the gear in Rack 1 is incestuously wired only among itself. Other racks go to DL connectors and interconnect as indicated. Rack 1 also receives I/O (input/output) from other racks.

WIRE RUN LIST 081301 DESCRIPTION (REV-1)	WIRE #	DL POS	IN FROM (SRC)	SRC CON	SRC B/U	OUT TO (DST)	DST CON	DST B/U	HRNS #	RAK #
RACK 1 > GEAR										
PCM-42 DLY IN	1	N/A	SBLD OUT 8	TRS [1/4]"	BAL	PCM-42 IN	TRS [1/4]	BAL	H-1	1
PCM-42 DLY OUT	2	N/A	PCM-42 OUT	TS [1/4]"	UNB	SBLD IN 8	TRS [1/4]	BAL	H-1	1
SANSAMP IN	3	N/A	SBLD OUT 6	TRS [1/4]"	BAL	SANSAMP IN	TRS [1/4]	BAL	H-1	1
SBLD SPARE IN R-1	4	N/A	FLTG TRS [1/4]"	TRS [1/4]"	BAL	SBLD IN 7	TRS [1/4]	BAL	H-1	1
SANSAMP OUT	5	N/A	SANSAMP OUT	XLR-F	BAL	SBLD IN 6	TRS [1/4]	BAL	H-1	1
SBLD SPARE OUT R-1	6	N/A	SBLD OUT 7	TRS [1/4]"	BAL	FLTG TRS [1/4]"	TRS [1/4]	BAL	H-1	1
D2 DLY IN L	7	N/A	SBLD OUT 11	TRS [1/4]"	BAL	D2 IN L	TRS [1/4]	BAL	H-1	1
D2 DLY IN R	8	N/A	SBLD OUT 12	TRS [1/4]"	BAL	D2 IN R	TRS [1/4]	BAL	H-1	1
D2 DLY OUT L	9	N/A	D2 OUT L	TRS [1/4]"	BAL	SBLD IN 11	TRS [1/4]	BAL	H-1	1
D2 DLY OUT R	10	N/A	D2 OUT R	TRS [1/4]"	BAL	SBLD IN 12	TRS [1/4]	BAL	H-1	1
RACK 2 > GEAR										
H-3000 IN L	11	1	DL-F3 (SBLD OUT 13)	DL-F3	BAL	H-3000 IN L	XLR-M	BAL	H-11	2
H-3000 IN R	12	2	DL-F3 (SBLD OUT 14)	DL-F3	BAL	H-3000 IN R	XLR-M	BAL	H-11	2
H-3000 OUT L	13	3	H-3000 OUT -	XLR-F	BAL	DL-F3 (SBLD IN13)	DL-F3	BAL	H-11	2
H-3000 OUT R	14	4	H-3000 OUT R	XLR-F	BAL	DL-F3 (SBLD IN14)	DL-F3	BAL	H-11	2

Figure 2.4.3 Samsrax wires tab.

- WIRE # – pretty self-evident: what number the wire is in the system. Sam's set-up has 110 wires in it, plus AC power cords. So what you're seeing here is the first 14 wires of a wire list that goes to 110 items. In olden days, each wire carried only the number, until we audio folk got smart and realized the utility of naming our wires as well as numbering them.
- DL POS – the wire's position in its associated DL plug, if it happens to go to one. Each wire in the DL has three contact points for its high/low/ground, and no DL in this system carries more than 12 wires. So by giving the wire's position in the DL, I know exactly which pins it's supposed to go to in that DL. Since both the Switchblade and the PCM-42 live in Rack 1, they don't go to a DL, and wire # 1 gets an 'NA' (not applicable) for the DL POS.
- IN FROM (SRC) – that's 'in from (source)' in shorthand. It's always confusing to remember that the output of one unit goes to the input of another unit. So each wire has (in a sense) two names. The wire called PCM-42 DLY IN is the same wire that comes from SBLD OUT 8, also known as Switchblade Out 8. Thus, the source (SRC) for the PCM-42 input is Output 8 of the Switchblade. If you're really cool, label the wire with an exact (differing) description at each end. As an example, the PCM-42 input wire would have two different labels. One label would say 'Wire #1/PCM-42 DLY IN' and go on the end connected to the PCM-42. The other end of the wire connected to Switchblade Out 8 would have a label that says 'Wire #1/SBLD OUT 8'. If you label in this manner, you will *always* know where *both* ends of any wire need to be connected. That saves oodles of time later on.
- SRC CON – the type of connector that needs to go *on the wire*, to mate properly with the unit the signal feed is coming from (the source). For the PCM-42, like many of the units in Sam's racks, the connectors are guitar plugs, so I decided to describe them as TRS {1/4} inch, stereo – tip/ring/sleeve, or TS {1/4} inch, mono – tip/sleeve. It's a lot faster to write than '{1/4} inch stereo/mono guitar plug'.
- SRC B/U – my telegraphic way of asking if the source of the signal is balanced or unbalanced. This makes a difference to the type of connector used and how it is wired.
- OUT TO (DST) – if you're catching on to my use of laconic abbreviations, you'll likely have guessed that this is 'out to (destination)' in condensed form. So this is where the signal is supposed to go to, to wind up, at the end of its circuitous journey. Our example of the PCM-42 delay shows us that Switchblade Output 8 terminates at the PCM-42 input. That's exactly where you'd expect, from our description of the wire as PCM-42 DLY IN.

- DST CON – or 'destination connector' again; this is the type of connector that goes *on the wire* to mate properly with the unit receiving the signal. For the PCM-42, that's a TRS {1/4} inch – if you don't know what that is, go back and read SRC CON again. Shame on you!
- DST B/U – once again, an abbreviated way of asking if the destination (this time, of the signal) is balanced or unbalanced. This (still) makes a difference in the type of connector used and how it is wired.
- HRNS # – it doesn't take too much effort to understand that this is 'harness number'. Remember the very beginning of Section 1? 'Wires are bundled together to become harnesses or cables.' Sam's racks have IIRC, 15 harnesses.
- RAK # – if you guessed 'rack number' you hit the jackpot! Sam's rig has three racks, and it's useful to know which rack a given wire lives in. Some wires are not in the racks at all, since the interconnect cables go between the racks, with a male DL connector on both ends. Each rack has female connectors that the male interconnect cables mate with – but only in season.

Whew, quite a list! And we're still not done – I just ran out of space on the page showing the spreadsheet layout. There's more! Remember, this spreadsheet was laid out to print on legal sized paper – 8.5 inch by 14 inch. So I have to continue on the next page to show what was cut off the right side of the spreadsheet (Figure 2.4.4).

WIRE RUN LIST 081301 DESCRIPTION (REV-1)	HRNS #	RAK #	NOTES
RACK 1>GEAR			
PCM-42 DLY IN	H-1	1	DIRECT SBLD I/O TO GEAR
PCM-42 DLY OUT	H-1	1	DIRECT SBLD I/O TO GEAR
SANSAMP IN	H-1	1	DIRECT SBLD I/O TO GEAR
SBLD SPARE IN R-1	H-1	1	DIRECT SBLD I/O TO GEAR
SANSAMP OUT	H-1	1	DIRECT SBLD I/O TO GEAR
SBLD SPARE OUT R-1	H-1	1	DIRECT SBLD I/O TO GEAR
D2 DLY IN L	H-1	1	DIRECT SBLD I/O TO GEAR
D2 DLY IN R	H-1	1	DIRECT SBLD I/O TO GEAR
D2 DLY OUT L	H-1	1	DIRECT SBLD I/O TO GEAR
D2 DLY OUT R	H-1	1	DIRECT SBLD I/O TO GEAR
RACK 2>GEAR			
H-3000 IN L	H-11	2	GEAR TO DL-F3
H-3000 IN R	H-11	2	GEAR TO DL-F3
H-3000 OUT L	H-11	2	GEAR TO DL-F3
H-3000 OUT R	H-11	2	GEAR TO DL-F3

Figure 2.4.4 Samsrax wires tab 2.

Notice that I've allowed an overlap of the DESCRIPTION, HRNS # and RAK # columns to help orient you in the spreadsheet. In reality, each column occurs just once.

The only part of this spreadsheet that was cut off is the NOTES. But notes can be very important! In our example of the PCM-42, the note reminds me that both the input and output of the PCM-42 go to/from the Switchblade directly – no DL connection is needed.

If you want the real flavor of this spreadsheet, either download and print the Excel file, or Xerox the pages with the spreadsheet on them, and paste/tape them together so you can see the completed 8.5 inch by 14 inch layout.

The individual wires weren't the only thing I tracked in Sam's spreadsheets. I also had a tab for a list of the gear itself, a tab for the harnesses, and a tab for purchased items with costs. This made it very easy to make notes about the gear, visualize the harnesses, and track expenses.

I'm going to show you (part of) each of these other tabs, to fire your imagination as to all the ways using a spreadsheet can help you organize your project. But I'm not going to discuss them – just show you the examples (Figures 2.4.5, 2.4.6 and 2.4.7). Otherwise this would become a section of encyclopedic proportions, and the AWG wouldn't fit in your tool kit.

I hope the examples above inspire you to create and use your own spreadsheets to organize projects – wiring and otherwise.

After using them for a while, you'll wonder how you ever did without them. And you'll realize how painful this process of organizing information was in the BODBC – that's in the 'Bad Old Days Before Computers'.

There are four mildly advanced features of spreadsheets that I want to describe for you.

First is the 'sum' function. For those of you unsure of the word, it means (in this case) to add up, to total. If I had room, I could show you that the COST column in the spreadsheet in Figure 2.4.7 is summed (added up) at the bottom of the column, to give me a grand total.

GEAR LIST RACK/GEAR	# RU	INPUT BAL/UNB	INFO #/CONN	IN FROM	OUTPUT BAL/UNB	INFO #/CONN	OUT TO	MIDI	AUD WIRE LENGTH	MID WIRE LENGTH
R1-PCM-42 DLY	1	1:BAL	1 XLR	SWBLD	1:UNB	1 TRS	SWBLD	NO	2 × 2'	NA
R1-SANSAMP	1	2:BAL	2 XLR	SWBLD	2:BAL	2 XLR	SWBLD	YES	2 × 2'	
R1-D2 DLY	1	2:BAL	2 XLR	SWBLD	2:BAL	2 XLR	SWBLD	YES	2 × 2'	
R1-SWBLD	1	NA	16:TRS	NA	NA	16:TRS	NA	YES		
SWBLD I/O 1	NA	NA	1 TRS	DL:MACK AUX OUT 1	NA	1 TRS	DL:MACK LN IN 5	NA		
SWBLD I/O 2	NA	NA	1 TRS	DL:MACK AUX OUT 2	NA	1 TRS	DL:MACK LN IN 6	NA		
SWBLD I/O 3	NA	NA	1 TRS	DL:MACK AUX OUT 3	NA	1 TRS	DL:MACK LN IN 7	NA		
SWBLD I/O 4	NA	NA	1 TRS	DL:MACK AUX OUT 4	NA	1 TRS	DL:MACK LN IN 8	NA		
SWBLD I/O 5	NA	NA	1 TS	DL:VOL PDL	NA	1 TS	DL:VOL PDL	NA		
SWBLD I/O 6	NA	NA	1 TRS	SANSAMP-L	NA	1 TRS	SANSAMP-L	NA		
SWBLD I/O 7	NA	NA	1 TRS	SANSAMP-R	NA	1 TRS	SANSAMP-R	NA		
SWBLD I/O 8	NA	NA	1 TRS	PCM-42	NA	1 TRS	PCM-42	NA		
SWBLD I/O 9	NA	NA	1 TRS	DL:PCM-70>1-L	NA	1 TRS	DL:PCM-70>1-L	NA		
SWBLD I/O 10	NA	NA	1 TRS	DL:PCM-70>1-R	NA	1 TRS	DL:PCM-70>1-R	NA		
SWBLD I/O 11	NA	NA	1 TRS	D2-L	NA	1 TRS	D2-L	NA		
SWBLD I/O 12	NA	NA	1 TRS	D2-R	NA	1 TRS	D2-R	NA		
SWBLD I/O 13	NA	NA	1 TRS	DL:H-3000-L	NA	1 TRS	DL:H-3000-L	NA		
SWBLD I/O 14	NA	NA	1 TRS	DL:H-3000-R	NA	1 TRS	DL:H-3000-R	NA		
SWBLD I/O 15	NA	NA	1 TRS	DL:PCM-70>2-L	NA	1 TRS	DL:PCM-70>2-L	NA		

Figure 2.4.5 Samsrax gear tab.

HARNESS DESCRIPTION	NAME	# WIRES	CONNECTOR # AND TYPE									HARDWARE
			TRS-M	XLR-M	XLR-F	(1/8)" TRS	RCA	DL-M	DL-F	MIDI-M	MIDI-F	
SWBLD <O>R-1 GEAR	H-1	8	11	5	4	0	0	0	0	0	0	
SWBLD <O>R-2 DL	H-2	8	8	4	4	0	0	0	1	0	0	MAP PLT
SWBLD <O>R-3 DL	H-3	12	12	0	0	0	0	0	1	0	0	MAP PLT
SWBLD>VOL PDL	H-4	2	4	0	0	0	0	0	0	0	0	
R-1 MIDI>SWBLD>MID SPLT	H-5	2	0	0	0	0	0	0	0	2	2	MAP PLT
R-2 MIDI >O PLT	H-6	2	0	0	0	0	0	0	0	2	2	MAP PLT
MID SPLT R-1	H-7	1	0	0	0	0	0	0	0	2	0	
MID SPLT R-2	H-8	2	0	0	0	0	0	0	0	4	0	
R-1>R-2 AUD INTERCONN	H-9	12	0	0	0	0	0	2	0	0	0	
R-1>R-3 AUD INTERCONN	H-10	12	0	0	0	0	0	2	0	0	0	
R-2>GEAR TO SWBLD DL	H-11	8	0	4	4	0	0	0	1	0	0	
R-3>EXT AUD INTERCONN	H-12	8	4	2	0	4	0	1	0	0	0	
R-3> MXR TO DL>R-?	H-13	12	12	0	0	0	0	0	1	0	0	
R-3>MXR TO EXT AUD DL	H-14	8	4	2	0	0	4	0	1	0	0	
TUNER A/D PWR	H-15	1	2	0	0	0	0	0	0	0	0	
TOTAL		98	57	17	12	4	4	5	5	10	4	
			16RA									
			4" STR									

Figure 2.4.6 Samsrax harness tab.

PURCHASE LIST						
DESCRIPTION	QTY	SOURCE	PHN	COST	DATE	PART #
250' ROLL MNSTR AUD	1	SAM ASH	212-586-1100	$312.50		MSL500B250
RIGHT ANGLE {1/4}" TRS	18	NEUTRIK		$90.00		NNP3RC
STRAIGHT {1/4}" TRS	45	NEUTRIK		$135.00		NNP3C
XLR-M CABLE	14	NEUTRIK		$42.00		NNC3MXB
XLR-F CABLE	9	NEUTRIK		$30.60		NNC3FXB
STRAIGHT {1/8}" TRS	3	CANARE		$12.00		CF12
RCA-M CABLE	5	CANARE		$13.75		CF9
DL-M	5	CANNON		$180.00		CDL96P
DL-M HOOD	5	CANNON		$98.00		E16230590
DL-M HANDLE	5	CANNON		$12.50		CDLH
DL-F	5	CANNON		$35.00		CDL96R
BAG 100 DL PINS	2	CANNON		$50.00		CDLPIN202
ROLL 62/36/2 SOLDER	1	N/A		$0.00		N/A
HEAT-SHRINK {3/64}"	2	COLEFLEX		$1.70		C221364
HEAT-SHRINK {1/8}"	2	COLEFLEX		$1.82		C22118
HEAT-SHRINK {3/16}"	2	COLEFLEX		$2.40		C221316
HEAT-SHRINK {1/4}"	2	COLEFLEX		$3.00		C22114
HEAT-SHRINK GUN	1	MASTER		$85.00		M10008
BAG 100 TIE-WRAPS	1	PANDUIT		$15.00		PPLT2IC
BAG TIE-WRAP MOUNTS	1	SAM ASH		$19.95		SCTMN
PK PANDUIT PLL-12-Y3	1	PANDUIT		$89.00		PPLL12Y3
DIN-M CABLE	6	SWITCHCRAFT		$18.00		S05BL5M
DIN-F CHASS	5	SWITCHCRAFT		$10.00		S57GB5F
20' MIDI WIRE (MOGMI)	1	MOGAMI		$6.80		M2944BK

Figure 2.4.7   Samsrax purchase tab.

The spreadsheet will do this automatically, which beats the heck out of doing it manually or with a calculator. The kindly spreadsheet will also *recalculate* the sum whenever you add or remove items – so you can begin to see how useful this feature can be.

How to invoke the 'sum' function varies slightly with each spreadsheet. Look in the 'Help' section of the spreadsheet you're using. It is often invoked by first highlighting a column of numbers (sum only works with numbers), and then clicking on a 'Σ' symbol. The sum (total) is then created at the bottom of the column of numbers.

Second is the 'show/hide' feature. This allows you to see and print only the columns and rows you need to see. In Excel, you first highlight *only* the rows/columns you want to hide. Then click on 'Format' in the menu bar at the top of the page, choose 'Rows' or 'Columns' and, finally, 'Hide' or 'Unhide'.

Download my example spreadsheets and play with this feature. A little practice and you'll soon be expert in the 'show/hide' feature.

The third function is called 'freeze/unfreeze panes' and is tragically misnamed. Much more accurate would be 'lock/unlock cells'. This function allows you to freeze (lock) the cells at both the top and the left side of the spreadsheet.

If you lay out your spreadsheet with your categories at the furthermost left and at the top – which is how the spreadsheet is designed to work – and then 'freeze' the categories, they will *remain* visible as you scroll through the hundreds, or thousands, of entries in your spreadsheet. This is *extremely* useful in keeping oriented within the vast sea of data you will accumulate!

In Excel, first place the cursor/selected cell *below*, or to the *right*, of the cells you want to freeze. You can do *both* at once, to freeze both topmost cells and leftmost cells. Then from the Menu bar at the top of the screen, select 'Window' and then 'freeze panes'. Follow the same procedure and 'unfreeze panes' to unlock the cells. Then, when you scroll in your spreadsheet, the leftmost and/or topmost cells will remain visible.

Gnumeric works a little differently: place the cursor as above, select 'View' from the Menu bar and then 'freeze/unfreeze panes' to achieve the same result.

Describing a dynamic function like 'freeze panes', is sort of like describing riding a bicycle ... Download my example spreadsheets and scroll within them – especially the Samsrax spreadsheet. The function and utility of freeze panes will soon be self-evident. Learn to freeze your own panes and become an ultra-cool spreadsheet creator!

The fourth (and last) function I'll describe is 'sort by'. Just as we can sum by a column of numbers, we can *sort* a whole sheet by any given column. And no, you cannot sort by rows. That's just how the pesky things work.

WIRE RUN LIST 081301 DESCRIPTION (REV-1)	WIRE #	DL POS	IN FROM (SRC)	SRC CON	SRC B/U
RACK 1>GEAR					
PCM-42 DLY IN	1	NA	SBLD OUT 8	TRS {1/4}"	BAL
PCM-42 DLY OUT	2	NA	PCM-42 OUT	TS {1/4}"	UNB
SANSAMP IN	3	NA	SBLD OUT 6	TRS {1/4}"	BAL
SBLD SPARE IN R-1	4	NA	FLTG TRS {1/4}"	TRS {1/4}"	BAL
SANSAMP OUT	5	NA	SANSAMP OUT	XLR-F	BAL
SBLD SPARE OUT R-1	6	NA	SBLD OUT 7	TRS {1/4}"	BAL
D2 DLY IN L	7	NA	SBLD OUT 11	TRS {1/4}"	BAL
D2 DLY IN R	8	NA	SBLD OUT 12	TRS {1/4}"	BAL
D2 DLY OUT L	9	NA	D2 OUT L	TRS {1/4}"	BAL
D2 DLY OUT R	10	NA	D2 OUT R	TRS {1/4}"	BAL
RACK 2>GEAR					
H-3000 IN L	11	1	DL-F3 (SBLD OUT 13)	DL-F3	BAL
H-3000 IN R	12	2	DL-F3 (SBLD OUT 14)	DL-F3	BAL
H-3000 OUT L	13	3	H-3000 OUT L	XLR-F	BAL
H-3000 OUT R	14	4	H-3000 OUT R	XLR-F	BAL
PCM-70/1 IN	15	5	DL-F3 (SBLD OUT 9)	DL-F3	BAL
SBLD SPARE IN R-2	16	6	DL-F3	DL-F3	BAL
PCM-70/1 OUT L	17	7	PCM-70/1 OUT L	TS {1/4}"	UNB
PCM-70/1 OUT R	18	8	PCM-70/1 OUT R	TS {1/4}"	UNB
PCM-70/2 IN	19	9	DL-F3 (SBLD OUT 15)	DL-F3	BAL
SBLD SPARE OUT R-2	20	10	DL-F3	DL-F3	BAL
PCM-70/2 OUT L	21	11	PCM-70/2 OUT L	TS {1/4}"	UNB
PCM-70/2 OUT R	22	12	PCM-70/2 OUT R	TS {1/4}"	UNB
RACK 2>SBLD (R-1)					
H-3000 IN L	23	1	DL-F1 (SBLD 13)	DL-F1	BAL
H-3000 IN R	24	2	DL-F1 (SBLD 14)	DL-F1	BAL
H-3000 OUT L	25	3	DL-F1 (SBLD 13)	DL-F1	BAL
H-3000 OUT R	26	4	DL-F1 (SBLD 14)	DL-F1	BAL

Figure 2.4.8   Samsrax wire tab – unsorted.

Let's take a Samsrax example again. First, I'll show you the unsorted form of a part of the 'WIRES' tab. Then I'll show you an ascending sort. I might use this feature to verify which wires go to DL and which ones don't.

Notice that the top two rows are *not* affected, as they are frozen more solid than the Arctic tundra.

The two examples show an unsorted (Figure 2.4.8) and a sorted view (Figure 2.4.9) of part of this tab. What is *not* visible is that the *whole* spreadsheet has been sorted by the IN FROM (SRC) column to assist me in determining if I have accounted for all the connections.

WIRE RUN LIST 081301 DESCRIPTION (REV-1)	WIRE #	DL POS	IN FROM (SRC)	SRC CON	SRC B/U
D2 DLY MIDI OUT	106	NA	D2 DLY MIDI OUT	DIN-M5	NA
D2 DLY OUT L	9	NA	D2 OUT L	TRS {1/4}"	BAL
D2 DLY OUT R	10	NA	D2 OUT R	TRS {1/4}"	BAL
SBLD SPARE OUT R-2	32	10	DL-F1	DL-F1	BAL
SBLD SPARE IN R-2	28	6	DL-F1	DL-F1	BAL
PCM-70/1 OUT R	30	8	DL-F1 (SBLD 10)	DL-F1	BAL
H-3000 IN L	23	1	DL-F1 (SBLD 13)	DL-F1	BAL
H-3000 OUT L	25	3	DL-F1 (SBLD 13)	DL-F1	BAL
H-3000 IN R	24	2	DL-F1 (SBLD 14)	DL-F1	BAL
H-3000 OUT R	26	4	DL-F1 (SBLD 14)	DL-F1	BAL
PCM-70/2 IN	31	9	DL-F1 (SBLD 15)	DL-F1	BAL
PCM-70/2 OUT L	33	11	DL-F1 (SBLD 15)	DL-F1	BAL
PCM-70/2 OUT R	34	12	DL-F1 (SBLD 16)	DL-F1	BAL
PCM-70/1 IN	27	5	DL-F1 (SBLD 9)	DL-F1	BAL
PCM-70/1 OUT L	29	7	DL-F1 (SBLD 9)	DL-F1	BAL
SBLD SPARE IN R-2	16	6	DL-F3	DL-F3	BAL
SBLD SPARE OUT R-2	20	10	DL-F3	DL-F3	BAL
H-3000 IN L	11	1	DL-F3 (SBLD OUT 13)	DL-F3	BAL
H-3000 IN R	12	2	DL-F3 (SBLD OUT 14)	DL-F3	BAL
PCM-70/2 IN	19	9	DL-F3 (SBLD OUT 15)	DL-F3	BAL
PCM-70/1 IN	15	5	DL-F3 (SBLD OUT 9)	DL-F3	BAL
MXR LN IN 9	45	9	DL-F4 (SBLD OUT 1)	DL-F4	BAL
MXR LN IN 10	46	10	DL-F4 (SBLD OUT 2)	DL-F4	BAL
MXR LN IN 11	47	11	DL-F4 (SBLD OUT 3)	DL-F4	BAL
MXR LN IN 12	48	12	DL-F4 (SBLD OUT 4)	DL-F4	BAL
MXR LN IN 7	43	7	DL-F4 (SPARE)	DL-F4	BAL
MXR LN IN 8	44	8	DL-F4 (SPARE)	DL-F4	BAL

Figure 2.4.9   Samsrax wire tab – sorted.

I hope this brief journey through the land of spreadsheets has left you enthused and eager to use them to assist you in your wiring endeavors.

There are whole books written about how to use different spreadsheets. I cannot condense all that information into one section. My goal is to show you how to use spreadsheets to assist you in wiring and motivate you to learn more. Download my examples and start playing!

2.5

Resources

This is a 'bonus' section that Focal Press allowed me to add at the last minute. Say 'Thank you' to the nice folks there.

A lot of the supplies and tools that are needed for wiring are not always easy to find locally. By having these URLs handy, you'll be able to find them with less effort. We'll also have them up on the AWG website that Focal has committed to, so you can copy and paste into your favorite web browser.

This list is in no way definitive, it's just a way to help you find what you need more easily.

Links from the AWG itself

I thought it would be easier to group all the links found in the AWG together, so you wouldn't have to leaf through all the sections to find the one you want.

Soldering iron tip cleaning products

- Radio Shack catalog no. 64-020 'Tip Tinner and Cleaner' (http://www.radioshack.com/product.asp?catalog%5Fname=CTLG&product%5 Fid= 64-020).

Other manufacturers also market a basically identical product:

- Multicore TTC1 (http://www.computronics.com.au/multicore/ttc/).
- A slightly different approach is the Apogee VTSTC – which also works (http://www.apogeekits.com/solder_tip_cleaner.htm).

Miller (or Miller-type) wire strippers

These are available from several sources:

- http://www.kelvin.com/Merchant2/merchant.mv?Screen=PROD&Product_ Code=520011&Category_Code=ELTOWS&Product_Count=5

- http://www.hmcelectronics.com/cgi-bin/scripts/product/5840-0004
- http://www.tecratools.com/pages/service/wirestrippers.html

Third Hand vise supplier (one of many)

The URL for Shor and the TH web page is http://shorinternational.com/TweezersSlide.htm.

Wire labels

- The Radio Shack version is catalog no. 278-1616 (http://www.radioshack.com/product.asp?catalog%5Fname=CTLG&category%5Fname=CTLG%5F011%5F010%5F008%5F001&product%5Fid=278%2D1616&site=search).
- A Panduit number for a slightly larger label (1 inch by 1.5 inch) is S100 × 150YDJ (http://www.panduit.com/products/Products2.asp?partNum=S100×150YDJ&param=361).

Silver solder (2%)

There are many suppliers. This is Radio Shack – the catalog number is 64-013 and the URL is http://www.radioshack.com/product.asp?catalog%5Fname=CTLG&category%5Fname=CTLG%5F011%5F009%5F007%5F001&product%5Fid=64%2D013&FGL=1-001

Equi=Tech links

Q: What if I can only afford BACP for some of my gear?

A: Excellent question! Take a look at this link – http://www.equitech.com/faq/somequip.html

- E = T home page – www.equitech.com
- List of E = T clients – http://www.equitech.com/ourclients/someclients.html
- Martin Glasband's article that I quote – http://www.equitech.com/articles/rep1.html
- Roger Nichols' article that I quote – http://www.equitech.com/articles/power.html

Freeware spreadsheets

- Open Office – www.openoffice.org/
- Gnumeric – www.gnome.org/projects/gnumeric/downloads.shtml

Caig

- Home page for Caig (makers of ProGold/DeOxit Gold and Flux Wash) – www.caig.com

Tips on BNC connectors

- http://www.l-com.com/content/Tips.aspx – look in the 'Coaxial' section.

Also look at:

- http://www.l-com.com/multimedia/tips/tip_75Ohm.pdf
- http://www.l-com.com/multimedia/tips/tip_50ohm.pdf

Links for tools/supplies/gear

These are links for tools, supplies and gear that are not included in the body of the AWG. They are in no particular order, except as I remembered them:

- Mogami wire (http://www.mogami.com/e/) – an excellent compromise between cost, quality and ease of wiring. Its spiral wrap shield makes it flexible and fast to work with.
- Monster cable (http://www.monstercable.com/) – might be the world's most hi-fi wire, but very expensive and slow to work with (braided shield):
- Canare wire (http://www.canare.com/index.cfm?objectid=2B02BB70-3048-7098-AF05D17CCDB6E617) – another excellent compromise, especially their star-quad wire. But like the Monster, star-quad is a braided shield wire.
- Mouser Electronics (http://www.mouser.com/index.cfm?handler=home) – parts, tools, wire; nice folks. They also have 2% silver solder if there's no Radio Shack near you.
- Newark Electronics (http://www.newark.com/) – more parts, tools, etc. Bigger, more stuff, less personal.
- Markertech (http://www.markertek.com/index.asp) – 60 000 oddball audio/video items that are hard to find.
- Techni-Tool (http://www.techni-tool.com/) – expensive, but very high quality tools.
- Xuron (http://www.xuron.com/) – makes some of the world's best hand tools. Their model 2175 flush-cutting dyke is one of my all-time favorites. Ultra-sharp, very strong, lasts forever. Buy one – you won't regret it.

- Dale Pro Audio (http://www.daleproaudio.com/) – been in business for decades, good folks. Mostly audio gear.
- Sweetwater Sound (http://www.sweetwater.com/) – among the better online audio stores. Mainly music/audio gear and accessories.

Reference links

Here are some links to sites that discuss concepts covered in the AWG:

- Star grounding (http://www.epanorama.net/documents/groundloop/rack_wiring.html) – somewhat flawed, but still educational. Read the docs on the Equi=Tech home site first.
- Ground loops (http://www.epanorama.net/documents/groundloop/) – ditto to the above (same author).
- Balanced AC power (http://www.epanorama.net/documents/groundloop/balanced_power.html) – same author as the star-ground link above. Not perfect, but still useful.
- Soldering techniques (http://www.epemag.wimborne.co.uk/solderfaq.htm) – excellent online reference.
- http://www.elexp.com/t_solder.htm – another good online soldering reference.
- The Wikipedia article on soldering (http://en.wikipedia.org/wiki/Soldering) – excellent and tons of links.

Remember, search engines are your friends – Google, Jeeves, Yahoo, Alta Vista, et al. They all use slightly different algorithms, so the same query will give different results on different engines. Use them all, until you find what you need.

Connector Modules

3.1 Guitar plug male connectors

The so-called 'guitar' plug is one of the oldest connectors still in use. With minor changes in form, you can see this type in use on your grandmother's pre-1900 telephone switchboard, and up to the present time.

Its large size makes it sturdy, if bulky, and it's a relatively easy connector to wire. The large surface area helps give good electrical contact, even when the plug's metal is slightly corroded. Small chunks of dirt and crud will often get knocked out of the way when the male plug is inserted in the female, so it's pretty roadworthy.

Standard guitar plugs are {1/4} inch in diameter. There's a miniature version, with a {1/8} inch diameter plug that's also much shorter in length. We show the sequence for that type in Section 3.3.

But both the standard and mini versions of the plug come in both stereo (two-conductor and shield) and mono (one-conductor and shield) versions. So I'm going to show you the wiring sequence for a stereo male guitar plug here.

My assumption is that it's easier for you to forget something than it is for you to remember it, right? So if you have a stereo male plug, just follow the sequence. If you have a *mono* plug, just *omit* the steps shown for the *low* conductor, as you will not have low, only high and shield connections. In other words, the stereo has high/low/shield, but the mono version is only high/shield.

To reinforce the above concepts, I've got a few quick pictures to show you, and then we can get on to the actual wiring of the plug.

Figure 3.1.1 is a close-up of the two solder tabs on a stereo male guitar plug. I've drawn two arrows to show exactly what part(s) I'm talking about. The longer part, that extends to the upper left in this picture, is both a strain relief for the wire and the part that the shield/drain gets soldered to. More on that later. Let's call the two tabs shown here the 'upper' and 'lower' tabs.

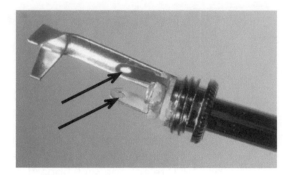

Figure 3.1.1   Solder tabs of stereo plug.

The lower tab goes to the center of the plug and down to the tip of the plug. It's the high/hot conductor. The upper tab goes to the ring of a stereo plug, but is *omitted* (not present) in a mono plug. It's the low/cold conductor. As a general rule: tip is high, ring is low, and the long barrel of the plug is used for drain/shield.

Since I want everyone to be totally clear on the difference between stereo and mono plugs, I've got a couple of side-by-side comparisons ready.

These pesky plugs are so shiny I had to put some white artist's tape behind the solder tabs, so you could see them against the shiny strain relief behind them (Figure 3.1.2). I hope it's all clear. On the left is a mono plug with one tab. On the right is a splendid example of a stereo plug with two solder tabs.

Figure 3.1.2   Mono/stereo comparison – 1.

Now that we're straight on the tabs, let's see the whole plug (Figure 3.1.3). Here we can see the business end(s) of our plug(s) – mono on the bottom and stereo on the top. Notice the ring on the stereo plug? That's the part the low conductor is connected to – and clearly omitted in the mono plug below it. So one tab=no ring, mono plug. Two tabs=has ring, stereo plug. All should now be clear as the finest crystal, and we can proceed to the actual sequence of wiring the thing.

Figure 3.1.3   Mono/stereo comparison – 2.

Wrap the wire firmly around your hand (Figure 3.1.4). This prevents the individual conductors from being pulled out when they are stripped and is only needed on short lengths of wire. If the wire is already harnessed or strain-reliefed, you don't have to wrap it around your hand.

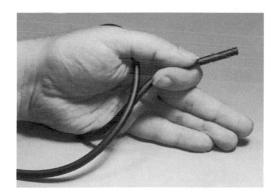

Figure 3.1.4   Wrap wire around hand.   **Figure 3.1.5**   Measure against plug.

Since different makes of plugs will have different lengths of the strain relief arm, you should measure the amount of outer jacket to be cut away against the plug itself (Figure 3.1.5). With this particular plug, we get a length to cut back of 1.25 inch. We're stripping back a little more than we need to, so we can have some slack to play with.

Note that by placing my thumbnail at the point being measured, I can guide my stripping tool safely to the exact spot I want. This makes it easier to strip back an accurate distance each time.

The wire I'm using for this example is two-conductor shielded Monster, one of the highest quality types available. Like many larger diameter wire types, it has a thick, rubbery outer jacket, which cuts more cleanly with a razor blade than with a pair of wire strippers. However, should you prefer, stripping the outer jacket off with wire strippers is also fine.

If using a blade, gently rock it at a right angle to the wire and then, using a light slicing motion, spin the blade around the wire, keeping it at a right angle at all times (Figure 3.1.6). This technique will create a super-clean cut of the outer jacket and, if done gently, will not harm the strands of the shield conductor that lie just inside the outer jacket.

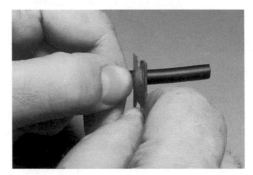

Figure 3.1.6   Cut outer jacket with razor blade.

Gently bend the cut in all directions to break away any slivers of the outer jacket that still connect both sides (Figure 3.1.7). If needed, cut the slivers lightly with the razor blade. The break should be complete on all sides.

Figure 3.1.7   Break open razor cut.

Lay the wire flat and cut from the circular break to the end of the wire, along the length of the wire (Figure 3.1.8). Do this delicately – this is not the time for force. Try to feel the wire as you cut, and let its resistance guide the strength of your slice.

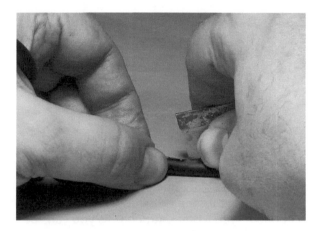

Figure 3.1.8   Cut along length of wire.

After you've finished the cut, peel away the outer jacket with your fingers. You shouldn't need any other tools.

Figure 3.1.9 shows the nature of the Monster wire we're using: it's a tinned and tightly braided shield – very effective and physically strong, but stiff, and a real pain in the fundament to work with. It also shows that we did the cut correctly; there are no broken strands in the shield conductor.

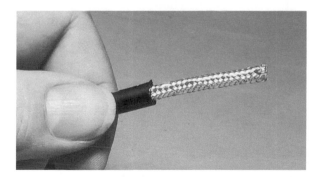

Figure 3.1.9   Outer jacket removed.

Push back the shield conductor strands away from the exposed end (Figure 3.1.10). This loosens the weave of the strands and makes them easier to unbraid.

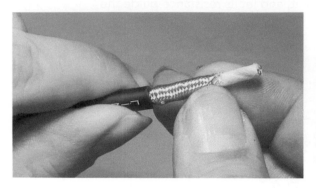

Figure 3.1.10   Push shield conductor strands back.   **Figure 3.1.11**   Unbraiding shield – 1.

Using any convenient small pointy object, carefully unbraid the shield strands (Figure 3.1.11). Do this one or two overlaps at a time; don't try to do a whole bunch at once – you'll break strands. As you can see here, the awl blade on a Swiss Army knife is ideal for this operation.

Keep going, keep going – it won't take long. Work all the braiding out of the shield strands, right down to where they break out from the outer jacket (Figure 3.1.12). This will help us a bit later, when we separate the wheat from the chaff, er, the shield strands from the leftover insulation underneath them.

Figure 3.1.12   Unbraiding shield – 2.

The shield strands are separated as in Figure 3.1.13. The alliteration is not deliberate – it's just what we're doing here. Gather all the shield strands off to one side and the fuzzy, frizzy insulation off in any other direction, as long as it's away from the shield, because we're about to do something calculatedly violent next.

Figure 3.1.13   Separating shield strands.

I've carefully gathered the shield, and both the high and low conductors, off to one side. I've also gathered the insulation off the other way and positioned it in the jaws of my dykes (Figure 3.1.14).

All is ready for the big cut! But before I do that, I'll take one final look to make sure all the conductors are out of harm's way. It's not hard to cut (or nick) one, and then you'd have all that unbraiding to do again. Very boring.

Figure 3.1.14   Cutting away insulation.

So look twice – heck, look three times. Cut once and cut right. You can always do a little clean-up work with the dykes after the main clump of insulation is gone.

Monster does their best to make quiet wire. So not only is the shield braided, the high and low conductors are twisted together, to reduce noise. Untangle them so you can deal with each conductor individually (Figure 3.1.15). Note the close-cropped collar of insulation? That's about

Figure 3.1.15   Untwisting conductors.

as close as you can get it with this type of wire.

As you can see in Figure 3.1.16, we've done the preliminary work properly. The (unbroken) shield strands are bunched to one side, the insulation is trimmed back neatly, and the low and high conductors are untangled, ready to be cut back.

Hold the wire in your hand (or a vise) and smooth out the strands of shield so they lie flat (Figure 3.1.17). The goal is to make a thin, broad surface, to solder onto the strain relief. Think of a strip of paper lying limply – not a twisted rope, as one might be tempted to make.

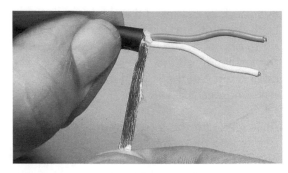

Figure 3.1.17   Smoothing out shield.

Figure 3.1.16   Measuring against connector.

Now that you've got those shield strands lying obediently flat, you need a trick to keep them that way. At the end of the shield strands, twist them around each other in a circular motion (Figure 3.1.18). This will leave the bulk of the strands lying flat, and flexible, but keep the whole bundle of strands together for soldering.

Figure 3.1.18   Twist shield at end.

Here, I only put a single twist in the shield conductors –
just enough to hold them together (Figure 3.1.19). The
arrow points to the twisted part. If you prefer, you can
keep going and twist them out to the end (toward the
right in the figure), as this part will be cut off later.

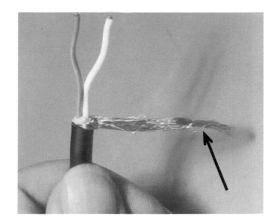

Solder the shield conductors at the twisted point (Figure
3.1.20). Use only enough solder to keep the strands
together at the twist. Don't let the solder run into the flat
portion you created – that would make the shield stiff,
when we need it to be flexible.

Put the wire aside for a bit and mount the connector in a
vise, with the strain relief facing up. Tin a broad area on
the arm of the strain relief, where the shield strands are
likely to wrap over it (Figure 3.1.21). Tin it lightly – just

Figure 3.1.19   Twist in shield.

enough to make a thin film on the surface for more solder to bond with.
A tasteful patina is required, not slobbery globs.

See the shiny surface and slight mounding in Figure 3.1.22? This is a
correctly tinned strain relief, ready for its final assembly. Be happy for it.

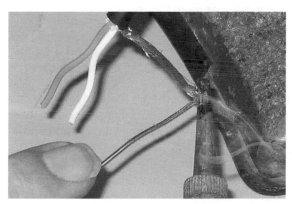

Figure 3.1.20   Solder twist in shield.

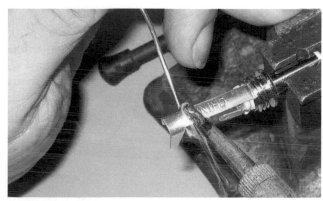

Figure 3.1.21   Tinning strain relief.

Figure 3.1.22   Correct tinning.

The next cuts are critical in length, so lay the wire against the connector to get a reality check on the needed length(s) (Figure 3.1.23). It's common that the optimum length for the high and low conductors is slightly different.

Figure 3.1.23   Sizing for cut-off.

If you're wondering what that mystical 'optimum length' is, here are some guidelines:

- The shield conductor must be positioned so it can be wrapped over the strain relief arm, past the 'U'-shaped yoke at the end, but not so close as to short out the high and low tabs.
- The high and low conductors must go in a slight arc (curve) to their respective tabs and be cut long, to allow enough stripped length for insertion in the holes in the center of each solder tab. The reason to have a slight arc is to make the conductor able to flex under stress – not snap off at the solder point.

In this case, the high conductor is {19/32} inch and the low conductor is {9/16} inch, but since connectors vary in length, so may your mileage.

See how it all lines up in Figure 3.1.24? The shield is past the yoke and I'm cutting the conductor I chose as high, to go slightly past the hole in the high solder tab.

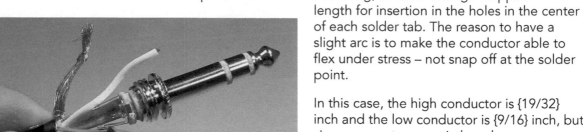

Figure 3.1.24   Cutting high conductor.

In this case, both the white and blue conductors are electrically identical. So you can choose either one for the low or high, but you *must* be consistent! If blue is high at one end of your cable, make darn sure it's high at the other end too. That was {19/32} inch, right?

The tabs on this connector are so close to the same length, I only need to make the low conductor {1/32} inch longer, for a total length of {19/32} inch when cut to fit (Figure 3.1.25). Could I get away with making both the low and high conductors the same length? Sure I could – but I want the insulation to go flush to the tab(s), so the extra length helps.

Figure 3.1.25   Cutting low conductor.

Strip both the high and low conductors back {1/16} inch (Figure 3.1.26). Why such a small amount? Because the insulation on Monster cable tends to shrink (or 'wick') back when the strands are heated to be tinned. You'll see this in the next steps.

Figure 3.1.26   Stripping both conductors.

Tin both the high and low conductors (Figure 3.1.27). Do this fairly quickly or the insulation will wick (melt) back too far.

Figure 3.1.27   Tinning the conductors.

Since our next step is to solder the connector, we *must* put the outer metal shell and inner plastic insulator onto the wire right now! If this step is forgotten, you get to *un*solder the tabs and try again.

You can see both those items in Figure 3.1.28. You lost the inner insulator? Improvise – use a bit of electrical tape after the soldering is complete. Gaffer tape, Scotch tape, even paper – you *must* provide some form of insulation between the tabs and the outer metal shell, or it'll be shorting time.

Figure 3.1.28   Putting on shell and insulator.

Position the wire so the shield is past the yoke and the strands of the low conductor are through the hole in the low solder tab (Figure 3.1.29). Try to get the insulation on the conductor to butt flush up to the tab, as it does here.

Figure 3.1.29   Inserting strands in tab.

Carefully crush (how's that for an oxymoron – but it's true) the strands down onto the solder tab to lie as flat as possible (Figure 3.1.30). This reduces the chance of shorting.

Soldering the low tab is the hard one (Figure 3.1.31). Once this solder point is done the wire will be locked into position. But right now, everything is still wiggly, so try and keep the strands aligned in the tab, and the shield draped over the strain relief.

Figure 3.1.30   Crush strands to tab.

Solder quickly, running in enough solder to fill the hole in the tab, and create a slight mounding (rounding) where the strands enter the tab hole.

And the white stuff in the figure isn't ectoplasm, it's a nice puff of rosin smoke. Dramatic, don't you think?

Figure 3.1.31   Solder low tab.

Figure 3.1.32   Completed low tab.

Figure 3.1.32 is a shot without the smoke, so you can see things clearly. Note the shiny solder, the mounding, and the insulation flush to the tab.

Try to make each solder point perfect; nothing is more important than the point you are working on right now – until the next one.

For the high tab, it helps to flip the connector around first, so the high tab is facing up. Then slip the strands of the high conductor into the hole of the high solder tab (Figure 3.1.33).

Next we have another crush job (Figure 3.1.34). Make the high strands lie flat against the high tab and chomp down with your pliers.

Figure 3.1.33   Insert high strands.

Figure 3.1.34   Crush high strands.

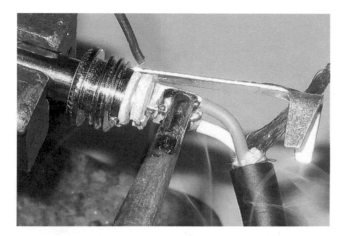

Figure 3.1.35   Solder high tab.

Solder the high tab with the same loving care as you did the low tab
(Figure 3.1.35). Solder quickly, running in enough solder to fill the hole in
the tab, and create a slight mounding (rounding) where the strands enter the
tab hole.

Figure 3.1.36 shows the construction details. See how flat the solder points
are? See how the insulation comes flush to the solder tabs? Nice – on to the
shield!

Flare the arms of the yoke with your pliers (Figure 3.1.37). This will reduce
the chance that contact with the yoke will melt the outer insulation jacket
when the shield strands are soldered.

Figure 3.1.36   High and low complete.

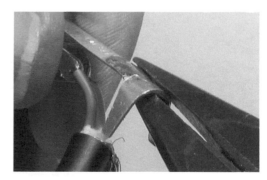

Figure 3.1.37   Spreading arms of yoke.

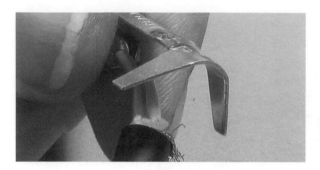

Figure 3.1.38   Properly flared yoke.

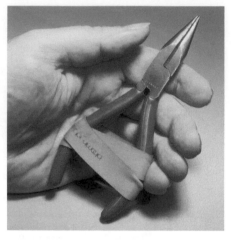

Spread the arms of that yoke nice
and wide (Figure 3.1.38). You'll crush
them down anyway, in the final
assembly.

Figure 3.1.39   Create weight for shield.

We need to make a weight to hang off the shield strands, one that will keep
them in position for soldering. Figure 3.1.39 shows a pair of pliers with a
rubber band for some clamping action – easy to improvise.

You can use anything that has sufficient weight to pull down on the shield
strands – hemostatic forceps, pliers or an old fishing sinker. Be creative – the
point where the weight is attached will be cut off the shield strands after
soldering anyway.

Rotate the connector in the vise so the strain relief is facing upward. Hold
the wire up against the flared yoke and tightly wrap the shield strands over
the arm of the strain relief (Figure 3.1.40).

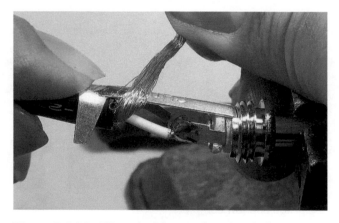

Figure 3.1.40   Wrap shield around strain relief.

Figure 3.1.41 Correctly positioned weight.

Our weight here is the pair of pliers hanging off the shield strands (Figure 3.1.41). See how tightly the strands are wrapped across the strain relief arm? Also note that the wire has fallen away from the yoke – which is a good thing. That way the yoke won't melt the outer jacket when we solder. All is in position for the last solder point to be done.

Press firmly down with the tip of the iron, for maximum heat transfer. Be liberal with the amount of solder you run in; here we want to fill all the voids in the shield strands, so they and the arm of the strain relief become one piece of metal (Figure 3.1.42). Just don't use so much solder as to cause blobs to appear on the shield strands.

The correct soldering of the shiny, silvery, shield strands is illustrated in Figure 3.1.43. OK, that alliteration is deliberate – but also totally accurate.

Figure 3.1.42 Soldering shield strands.

Figure 3.1.43 Correctly soldered shield.

Use dykes to cut away the excess shield strands (Figure 3.1.44). Cut as close as possible to the arm of the strain relief. Also cut at an angle, so the cut edge is beveled (angled) in relation to the strain relief arm. It gives you less to file smooth in the next step.

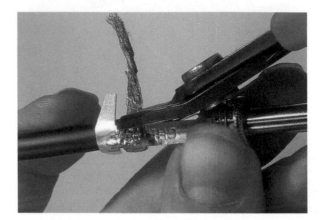

Figure 3.1.44   Cutting excess shield strands.

File the edge you just cut, with any size file that will fit. In Figure 3.1.45, I'm using about the biggest file that will fit the work. It was handy and will make short shrift of any rough edges.

Figure 3.1.45   Filing cut edge.

Figure 3.1.46 shows the correctly filed edge. Now is also a good time to cut away any stray strands that may have escaped your clutches, clean up any extra insulation, and generally tidy things up.

Figure 3.1.46   Correctly filed edge.

Here, our goal is to crimp and curl the strain relief yoke tightly around the curve of the wire. So bend it a little at a time, and work it so as to press it into the outer jacket evenly at all points. We want an arc that matches the wire, so do one side at a time. Figure 3.1.47 shows the first part of the crimp.

Figure 3.1.47   Crimping yoke – 1.

The second half of the operation is illustrated in Figure 3.1.48. We can see the curled yoke arm we just formed. Now we want to curl the other yoke arm over the first one, making a very strong, secure strain relief.

Figure 3.1.48   Crimping yoke – 2.

If done right, this combination of the support from the yoke arms, combined with the shield being soldered down, will make the strain relief as strong as the wire itself. Pretty cool, huh?

All the hard work is done, we just have to screw the outer shell in place and squirt the connector with our special goose grease – er, ProGold contact enhancer, that is. But I thought you might like to see the final product (Figure 3.1.49) before the outer shell hides all the details.

Figure 3.1.49   (Almost) complete connector.

Screw the outer metal shell firmly into place (Figure 3.1.50). And now, for the very last step.

Figure 3.1.51 shows the tip, ring and sleeve being sprayed with enhancer. I'm using Caig ProGold G5 and recommend that you do too! It prevents aerobic corrosion – which degrades conductivity. It improves electrical contact on all metals used in electrical connections, and is especially good for low-power connections like guitar outputs and mating stage boxes. Most guitars have very low power output, down in the microvolt range.

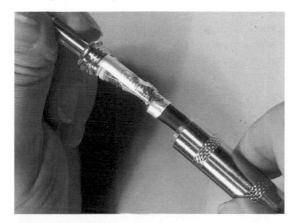

Figure 3.1.50   Screwing down outer shell.

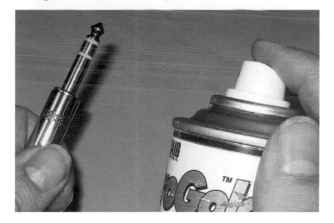

Figure 3.1.51   Spraying with ProGold.

At such a low level, any slight reduction in conductivity can be much more harmful then it would be at line level or speaker level. At higher power levels, the signal can cut through layers of (semi-conductive) corrosion.

Note: Caig Labs has recently changed the name from ProGold to DeoxIT Gold. It's the same stuff. They also have some new products that are interesting. More info about ProGold is on the Caig Labs website (www.caig.com).

That's it, folks. 'That's all she wrote' – or maybe I should say 'wired'. Now it's time to whip out that VOM you bought after I suggested you do so in Section 1. If you really read the whole book, you'll recall a mini-course in how to use it (the VOM) that was part of Section 2. You'll even remember what VOM, or DVOM, stands for.

Test your work, check it for shorts and high-resistance solder points. If it all checks out, enjoy your new connections!

Guitar plug female connectors

For those of you reading through the connector modules sequentially, a good part of the information on the male guitar plug is duplicated here. This is neither accidental nor laziness on my part. Rather, they are connectors in the same family, and each module is written as a 'stand-alone', to reduce flipping back and forth for instructions. So now for a little recycling, followed quickly by some new pictures.

The so-called 'guitar' plug is one of the oldest connectors still in use. With minor changes in form, you can see this type in use on your grandmother's pre-1900 telephone switchboard, and up to the present time.

Its large size makes it sturdy, if bulky, and it's a relatively easy connector to wire. The large surface area helps give good electrical contact, even when the plug's metal is slightly corroded. Small chunks of dirt and crud will often get knocked out of the way when the male plug is inserted in the female, so it's pretty roadworthy.

Standard guitar plugs are {1/4} inch in diameter. There's a miniature version, with a {1/8} inch diameter plug that's also much shorter in length. We show the sequence for that type in Section 3.3.

But both the standard and mini versions of the plug come in both stereo (two-conductor and shield) and mono (one-conductor and shield) versions. So I'm going to show you the wiring sequence for a stereo female guitar plug here.

My assumption is that it's easier for you to forget something than it is for you to remember it, right? So if you have a stereo female plug, just follow the sequence. If you have a *mono* plug, just *omit* the steps shown for the *low* conductor, as you will not have low, only high and shield connections. In other words, the stereo has high/low/shield, but the mono version is only high/shield.

To reinforce the above concepts, I've got a few quick pictures to show you, starting with the female plug in Figure 3.2.1, and then we can get on to the actual wiring of the plug.

Figure 3.2.1   Female plug.

How can you tell if I'm holding a mono or a stereo female guitar plug? Well, one way is to plug a male into it. If you get one click, it's mono, get two clicks and it's stereo – unless the contacts are mushy and fool you. But the easier way is to take the outer shell off and look at the solder tabs inside. Let's do that now in this next series of figures.

Figure 3.2.2 is an overall shot of a mono plug (at the front) and a stereo plug (at the back). See the difference in construction? The mono plug has only one solder tab, the stereo plug has two. This is shown in greater detail in Figure 3.2.3. This ultra-close-up clearly shows the one vs. two solder tab construction on the mono (left) and the stereo (right) versions of our beloved female guitar plug. Without her, there would be no headphone extensions – tragic.

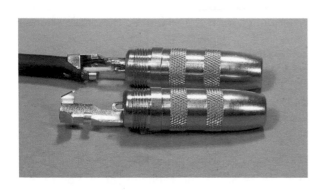

Figure 3.2.2   Stereo/mono plugs – 1.

Figure 3.2.3   Stereo/mono plugs – 2.

The hoopy looking thing on the left side of both plugs is a great place to solder the shield/drain wire. It's better *not* to crush it shut, as that makes the plug hard to recycle. The right-hand plug shows the drain soldered in place; the left-hand plug shows the bare drain solder tab.

Figure 3.2.4 is just to give you an idea of the dimensions involved. There's not a lot of room inside one of these babies, so your craftspersonship must be exemplary. Yeah, I know that's awkward – how about craftsoneship? Much, if not most, of the world's wiring is done by women.

Wrap the wire firmly around your hand (Figure 3.2.5). This prevents the individual conductors from being pulled out when they are stripped, and is only needed on short lengths of wire. If the wire is already harnessed or strain-reliefed, you don't have to wrap it around your hand.

In Figure 3.2.4, we saw that the soldering end of the plug was less than {3/4} inch. So with this plug, we can cut back a little more than 1.25 inch (Figure 3.2.6). We're stripping back a little more then we need to, so we can have some slack to play with. As long as you have extra, the length is not critical.

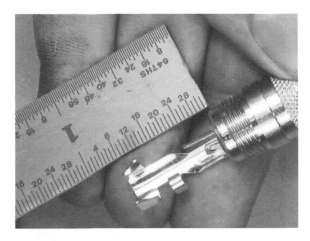

Figure 3.2.4 Dimensions.

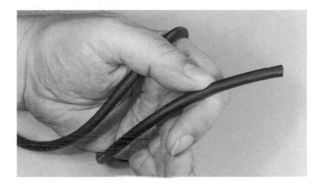

Figure 3.2.5 Wrap wire around hand.

Figure 3.2.6 Measure strip.

Note that by placing my thumbnail at the point being measured, I can guide my stripping tool safely to the exact spot I want. This makes it easier to strip back an accurate distance each time.

The wire I'm using for this example is two-conductor shielded Mogami, one of the highest quality types available. Like many larger diameter wire types, it has a thick, rubbery outer jacket, which cuts more cleanly with a razor blade than with a pair of wire strippers (Figure 3.2.7). However, should you prefer, stripping the outer jacket off with wire strippers is also fine.

Figure 3.2.7   Cut outer jacket with razor blade.

If using a blade, gently rock it at a right angle to the wire and then, using a light slicing motion, spin the blade around the wire, keeping it at a right angle at all times. This technique will create a super-clean cut of the outer jacket and, if done gently, will not harm the strands of the shield conductor that lie just inside the outer jacket.

Gently bend the cut in all directions to break away any slivers of the outer jacket that still connect both sides (Figure 3.2.8). If needed, cut the slivers lightly with the razor blade. The break should be complete on all sides.

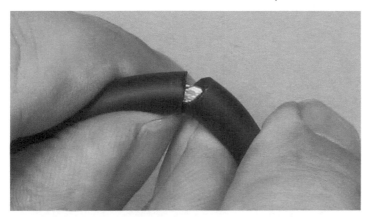

Figure 3.2.8   Break open razor cut.

This particular wire is very supple, and the outer insulation jacket can just be pulled off (Figure 3.2.9). One of my crude arrows helps illustrate the direction to pull. If the outer jacket is sticking to the shield, then you have to cut it lengthwise (see the details in Figure 3.1.8). I'm breaking my own rule about flipping – but this wire behaved itself and didn't need the additional cut.

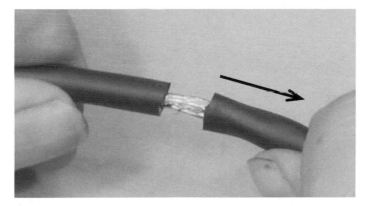

Figure 3.2.9   Pull end off.

This particular Mogami shield (Figure 3.2.10) is not tinned and is loosely woven, so it's flexible and easy to unbraid, unlike some other shields around here. I was good – I didn't press too hard with the razor blade – so none of the strands are broken.

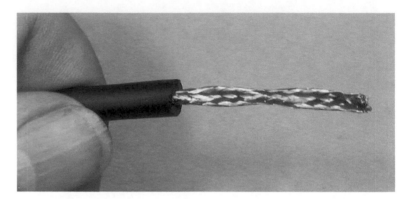

Figure 3.2.10   Detail of shield.

Push back the shield conductor strands away from the exposed end (Figure 3.2.11). This loosens the weave of the strands and makes them easier to unbraid.

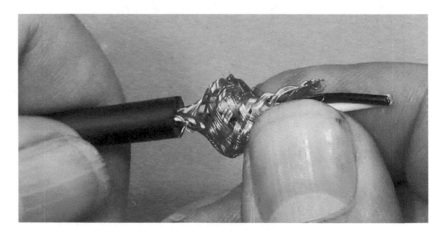

Figure 3.2.11   Push shield conductor strands back.

Using any convenient small, pointy object, carefully unbraid the shield strands (Figure 3.2.12). Do this one or two overlaps at a time; don't try to do a whole bunch at once – you'll break strands. As you can see here, the awl blade on a Swiss Army knife is ideal for this operation.

Keep going, keep going – it won't take long. Work all the braiding out of the shield strands, right down to where they break out from the outer jacket (Figure 3.2.13). This particular wire is very civilized, no fuzz, string, plastic wrapper, just the three conductors – nice! You might not be so lucky – be prepared for a little clean-up. Trim all leftover insulation as tightly as possible, without harm to the conductors.

Figure 3.2.12   Unbraiding shield – 1.

Figure 3.2.13   Unbraiding shield – 2.

Hold the wire in your hand (or a vise) and smooth out the strands of shield so they lie flat (Figure 3.2.14). The goal is to make a thin, broad surface, to solder onto the strain relief. Think of a strip of paper lying limply – not a twisted rope, as one might be tempted to make.

Figure 3.2.14   Smoothing out shield.

Now that you've got those shield strands lying obediently flat, you need a trick to keep them that way. At the end of the shield strands, twist them around each other in a circular motion (Figure 3.2.15). This will leave the bulk of the strands lying flat, and flexible, but keep the whole bundle of strands together for soldering.

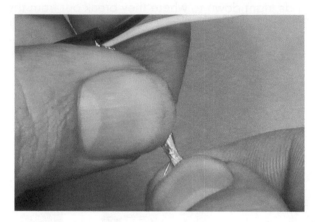

Figure 3.2.15   Twist shield at end.

Figure 3.2.16 is a fly's eye view of the twist at the end of the shield. Now we need some way to keep it twisted – which brings us to Figure 3.2.17.

Solder the shield conductors at the twisted point. Use only enough solder to keep the strands together at the twist. Don't let the solder run into the flat portion you created – that would make the shield stiff, when we need it to be flexible.

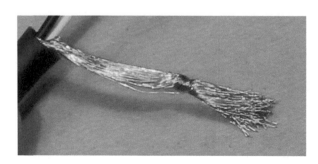

Figure 3.2.16   Twist shield detail.

Figure 3.2.17   Solder twist in shield.

The next cuts are critical in length, so lay the wire against the connector to get a reality check on the needed length(s) (Figure 3.2.18). It's common that the optimum length for the high and low conductors is slightly different.

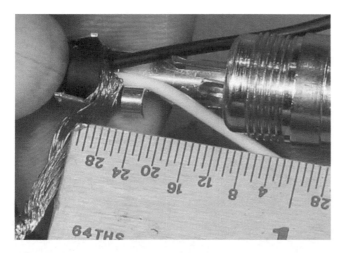

Figure 3.2.18   Sizing for cut-off.

In this case, both the white and black conductors are electrically identical. So you can choose either one for the low or high, but you *must* be consistent! If black is high at one end of your cable, make darn sure it's high at the other end too. Electrical convention in AC wiring is that black is low, but you're not bound by law to follow that – only to be consistent in your work.

I've laid a ruler against the work for comparison, but the plug itself is the place to determine measurements from, as each type will be slightly different.

If you're wondering what that mystical 'optimum length' is, here are some guidelines:

- The shield conductor must be positioned so it can be wrapped over the strain relief arm or, in this case, inserted in a solder tab, past the 'U'-shaped yoke at the end, but not so close as to short out the high and low tabs.

- The high and low conductors must go in a slight arc (curve) to their respective tabs, and be cut long, to allow enough stripped length for insertion in the holes in the center of each solder tab. The reason to have a slight arc is to make the conductor able to flex under stress – not snap off at the solder point.

In this case, the high conductor length is {9/16} inch and the low conductor is {7/16} inch, but since connectors vary in length, so may your mileage.

Make sure the outer jacket of the wire is correctly positioned in the strain relief yoke. Look just to the right of my left thumbnail in Figure 3.2.19 to see what I mean. Then cut the high conductor just a tiny bit longer than you think is needed. It's easy to cut back more – but hard to replace once cut.

Figure 3.2.19 Cutting high conductor.

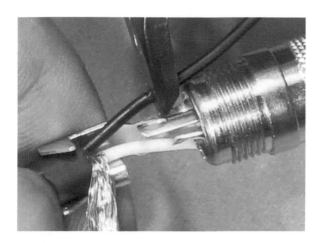

Figure 3.2.20   Cutting low conductor.

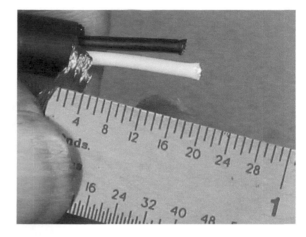

Figure 3.2.21   Finished cut length.

We have the same situation in Figure 3.2.20. Keep the outer jacket in position and cut the low conductor just a bit longer than you think is needed. Remember that arc we talked about a few shots back?

You can see from Figure 3.2.21 that I was telling the truth. The high conductor length really is {9/16} inch and the low conductor is {7/16} inch. That's what works for this particular connector, with this type of wire. Now for some stripping on the high and low conductors.

Strip both the high and low conductors back 1/16 inch (Figures 3.2.22 and 3.2.23). Why such a small amount? Because the insulation on cable tends to shrink (or 'wick') back when the strands are heated to be tinned. You'll see this in the next steps.

Figure 3.2.22   Stripping high conductor.

Figure 3.2.23   Stripping low conductor.

The finished strip of both insulated conductors is shown in Figure 3.2.24. The black is stripped {1/16} inch, the white is just a hair longer – that's OK too.

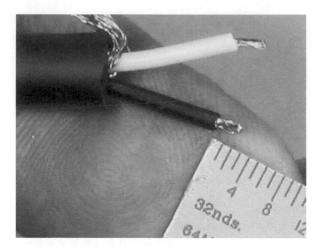

Figure 3.2.24   Finished strip.

OK, I skipped a shot. I should have shown tinning the high conductor – but I got too excited. So imagine really hard that you saw me do it and look at the shot of me tinning the low conductor (Figure 3.2.25).

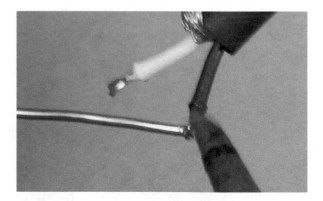

Figure 3.2.25   Tin both conductors.

We need to use a technique called 'beading' here, where the tinned ends of the conductors are loaded with extra solder. This forms (ideally) a grape- or bead-like shape – hence the name. This technique is used where it is not possible to feed the solder onto the heated conductor. Instead, the conductor itself must carry the solder as a 'payload'.

The first step in beading is to tin the end of the conductor. Sometimes it is possible to add the full payload of solder all at once. Often, however, it helps to let the tinning cool down and then add the beading (build-up) as a second step. Figures 3.2.26 and 3.2.27 show this process.

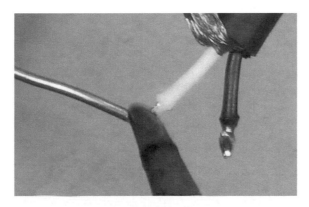

Figure 3.2.26   Beading high conductor – 1.

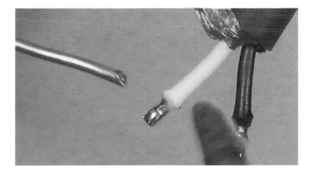

Figure 3.2.27   Beading high conductor – 2.

I got lucky when I tinned the low conductor and got a good payload of solder on it. It's more conical than grape shaped, but a small snip with my cutters will blunt the tip if needed. So in Figure 3.2.26 I'm loading up the high conductor with solder.

There's almost enough solder on the high conductor, so I'll lightly add just a bit more here.

Our artfully beaded conductors are shown in Figure 3.2.28. The conehead on the low will melt into position, and the bead on the high conductor is a perfect example of the technique.

Figure 3.2.28   Completed beading.

Figure 3.2.29   Putting on shell.

Since our next step is to solder the connector, we *must* put the outer metal shell and (if included) the inner plastic insulator onto the wire right now (Figure 3.2.29)! If this step is forgotten, you get to *un*solder the tabs and try again.

This plug didn't come with an inner insulator. Maybe you lost the inner insulator? Improvise – use a bit of electrical tape after the soldering is complete. Gaffer tape, Scotch tape, even paper – you *must* provide some form of insulation between the tabs and the outer metal shell, or it'll be shorting time.

Put the wire and the outer shell aside for awhile. Time now to play with the connector.

When it comes from the factory, the solder tabs are too close together – likely to short out. I'm flaring them out a bit in Figure 3.2.30 to get a tad more space. This is more visible in Figure 3.2.31.

Figure 3.2.30   Flaring solder tabs – 1.

Figure 3.2.31   Flaring solder tabs – 2.

I put a bit of white tape across the strain relief yoke so you could see the flared solder tabs more clearly. As you can see, I only bent them a little bit. Just enough for some breathing room – and maybe a bit of mylar electrical tape!

Space is so tight inside the connector that I want to pre-tin the high and low conductor solder tabs. This way, when I bring the solder-laden conductor into position against the solder-encrusted tab, they will bond immediately when heated. That's the premise, at least. In Figure 3.2.32, I'm tinning the high tab.

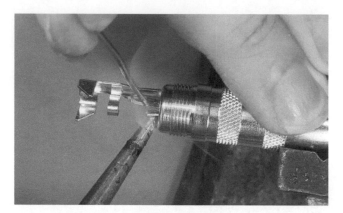

Figure 3.2.32   Tinning high tab.

Tin both tabs the same way – with a thick film of solder. Enough to provide 'wetting' action against the solder on the conductors, but not so much as to drip down and short.

In Figure 3.2.33 notice how I'm bracing my hand against the vise to stabilize my solder feed? This is like the brace you take for shooting pool.

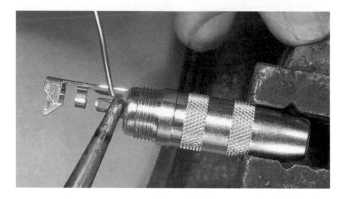

Figure 3.2.33   Tinning low tab.

Put the high conductor in position against the high conductor solder tab (Figure 3.2.34). Grip the wire in such a way that you will not be burned when bringing the soldering iron into position.

Figure 3.2.34   Placing high conductor.

Press the iron's tip firmly against the solder tab for maximum heat transfer (Figure 3.2.35). With thicker tabs, you may have to heat the solder on the conductor itself. Experiment, and see what works best for the connectors you have.

Figure 3.2.35   Soldering high conductor.

Once the solder has melted, hold the iron in place for no more than 1–2 seconds and then remove it. Allow the solder to cool without moving the conductor.

Once the high conductor is in place, it's easier to do the low one. In Figure 3.2.36 I show the low one pushed into position and in the act of being soldered. It's the same drill as the high tab – heat until molten, remove the iron, and allow to cool without moving the conductor.

Figure 3.2.36   Soldering low conductor.

I've used that white tape again to clearly show the spacing of the finished conductors in Figure 3.2.37. Notice that they are in no danger of shorting to each other, and both will still be a safe distance from the inside of the outer shell once it is screwed in place. By building it this way, the connector can be safely used with no additional insulation.

Figure 3.2.37   Completed conductors.

Remember that hoopy thing I showed you in Figure 3.2.3? Now is the time to put the drain conductor in it. Make sure the outer jacket is shoved well forward against the clamping tabs at the end of the strain relief. In Figure 3.2.38 one of the tabs is above my thumbnail, the other is hidden by the outer jacket.

Figure 3.2.38   Placing drain/shield.

Pardon my blurry solder in Figure 3.2.39 – I was trying to move it along the tab. This is one place where you can be generous with solder, filling in all the nooks and crannies in the shield/drain conductor. Just don't get carried away and create dripping blobs of molten solder. Use enough to fill the drain cup completely, but no more.

Figure 3.2.39   Soldering drain tab.

Cut away the excess drain conductor (Figure 3.2.40). File any sharp edges smooth.

The completed cut on the drain conductor is shown in Figure 3.2.41. It's time now to compress the strain relief tabs.

Figure 3.2.40   Cut off excess drain – 1.

Figure 3.2.41   Cut off excess drain – 2.

Here, our goal is to crimp and curl the strain relief yoke tightly around the curve of the wire. So bend it a little at a time, and work it so as to press it into the outer jacket evenly at all points. We want an arc that matches the wire, so do one side at a time. Figure 3.2.42 shows the first part of the crimp.

Figure 3.2.42   Crimping yoke – 1.

The second half of the operation is illustrated in Figure 3.2.43. We can see the curled yoke arm we just formed. Now we want to curl the other yoke arm over the outer jacket, making a very strong, secure strain relief.

Figure 3.2.43   Crimping yoke – 2.

If done right, this combination of the support from the yoke arms, along with the shield being soldered down, will make the strain relief as strong as the wire itself. Pretty cool, huh?

Screw the outer metal shell firmly into place (Figure 3.2.44). And now for the very last step.

Figure 3.2.44   Screwing down outer shell.

Figure 3.2.45   Spraying with ProGold.

Figure 3.2.45 shows the inside of the connector being sprayed with enhancer. Wipe off any overspray. I'm using Caig ProGold G5 and recommend that you do too! It prevents aerobic corrosion – which degrades conductivity. It improves electrical contact on all metals used in electrical connections, and is especially good for low-power connections like guitar outputs and mating stage boxes. Most guitars have very low power output – down in the microvolt range.

At such a low level, any slight reduction in conductivity can be much more harmful than it would be at line level or speaker level. At higher power levels, the signal can cut through layers of (semi-conductive) corrosion.

Note: Caig Labs has recently changed the name from ProGold to DeoxIT Gold. It's the same stuff. They also have some new products that are interesting. More info about ProGold is on the Caig Labs website (www.caig.com).

That's it, folks. 'That's all she wrote' – or maybe I should say 'wired'. Now it's time to whip out that VOM you bought after I suggested you do so in Section 1. If you really read the whole book, you'll recall a mini-course in how to use it (the VOM) that was part of Section 2. You'll even remember what VOM, or DVOM, stands for.

Test your work, check it for shorts and high-resistance solder points. If it all checks out, enjoy your new connections!

Mini-male guitar plug connectors

That's an awfully long name for such a small, short plug. For the sake of brevity (and mutual sanity), I'll abbreviate it as MMGP. These miniature versions of a standard {1/4} inch guitar plug are a perfect example of a 'second generation' connector. And if I happen to mention an MFGP, that's the female of the species.

First came the regular guitar plugs – the kind you'd stick into your, uh, electric guitar or your synthesizer. They were OK for places where you had plenty of room. But what about the headphone jack on your Walkman? Or the back of your computer for the audio ins and outs?

Nothing is ever 100% good – and the amazing, shrinking guitar plug was/is no exception. Its smaller size makes for a smaller electrical contact area, a more fragile connector, and one that's a lot more difficult to wire. The small size makes tolerances very tight for your work.

Despite this, it turned out that there were a lot of places where a smaller connector would work better. So, faster than you can say 'miniaturization', a smaller version was created. In fact, two different types were (and are) made.

Just like their big {1/4} inch brothers, MMGPs come in mono (which is a high and shield conductor type) and stereo (which has a high conductor, a low conductor and the shield). The same applies for MFGPs.

My dilemma is to show you as much as possible in the minimum amount of time, text and pictures. So I'm going to show you the wiring sequence for a stereo MMGP here.

My assumption is that it's easier for you to forget something than it is for you to remember it, right? So if you have a stereo MMGP, just follow the sequence. If you have a *mono* MMGP, just *omit* the steps shown for the *low* conductor, as you will not have low, only high and shield, connections. In other words, the stereo has high/low/shield, but the mono version is only high/shield.

I'm also going to economize and show you only an MMGP, but not an MFGP, wiring sequence. Why? Other than the fact that I'm running out of room, the MFGP is not too commonly field wired. Typically, you wire MMGPs to insert into MFGP jacks on equipment. Female jacks? Shouldn't they be jills? Maybe – but they're not.

If you get stuck wiring an unlikely MFGP, look at the {1/4} inch version and mentally scale things down about 50%. Think small and delicate – that's half the battle.

To reinforce the above concepts, take a look at Figure 3.3.1, and then we can get on to the actual wiring of the plug.

Figure 3.3.1   Solder tabs of a stereo plug.

The figure is a close-up of the two solder tabs on a stereo MMGP. I've drawn two arrows to show exactly what part(s) I'm talking about. The longer part, which extends to the upper left in this picture, is both a strain relief for the wire and the part that the shield/drain gets soldered to. More on that later.

Let's call the two tabs I show the 'upper' and 'lower' tabs. Of course, if you turn the picture around, they would not remain so. But I have to differentiate them somehow.

The lower tab goes to the center of the plug and down to the tip of the plug. It's the high/hot conductor. The upper tab goes to the ring of a stereo plug, but is *omitted* (not present) in a mono plug. It's the low/cold conductor. As a general rule, tip is high, ring is low and the long barrel of the plug is used for drain/shield.

In Figure 3.3.1 you can also see the round black metal barrel of the MMGP. The especially aware reader will note the lack of an insulating sleeve (typically plastic or paper), to go between the metal barrel and the equally conductive metal of the plug, and your solder points. The cause? For some reason – poor design or packing error – this otherwise well-designed MMGP did *not* come with an insulating sleeve. I'll show you how to fix that later on.

Small plugs need to be soldered with small wire. The thick, heavy insulation of the wire I used for the {1/4} inch guitar plug won't work here – there is no room, even for a genius-level wireperson. So I've chosen some smaller, two-conductor, foil-shielded wire to work with – typical studio installation wire. It's West Penn type 291 to be totally specific.

Spiral strand-shielded wire, like Mogami or Gotham, would also work well here. Braided strand-shielded wire is best avoided. It has excellent strength and shielding properties, but is very slow to work with, as the braid must be unwoven a few strands at a time.

Wrap the wire firmly around your hand (Figure 3.3.2). This prevents the individual conductors from being pulled out when they are stripped, and is only needed on short lengths of wire. If the wire is already harnessed or strain-reliefed, you don't have to wrap it around your hand.

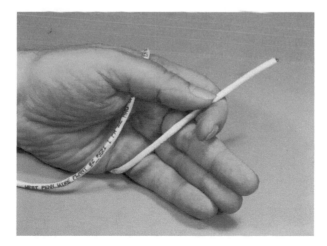

Figure 3.3.2   Wrap wire around hand.

Measure the wire a convenient distance back. Figure 3.3.3 shows the wire being measured against a single-edge razor blade, which gives a length of about {3/4} inch, but any repeatable length up to about 1 inch will do. It is easier to dress the wire correctly if you strip it longer than needed and then cut off the excess.

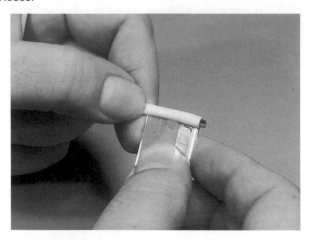

Figure 3.3.3   Measure for strip-off.

Note that by placing my thumbnail at the point being measured, I can guide my stripping tool safely to the exact spot I want. This makes it easier to strip back an accurate distance each time.

Even though I used a razor blade to judge the length of the cutback, I'll use a pair of strippers to cut the outer jacket insulation this time (Figure 3.3.4).

Figure 3.3.4   Cutting outer jacket with strippers.

Both a razor blade and strippers work well. Use whichever one you're most comfortable with.

This is the first part of the strip-off procedure. Cut into the outer insulation jacket enough to score it deeply, but not so hard as to nick the inner conductors. This can be done by 'feel' after you've done a few dozen.

Even better, set the guide on the strippers to the exact depth needed, using scrap wire, before you start any real work. If you're using strippers, I'm going to ask you to move them over to a part of the outer jacket closer to the edge, before pulling on the outer jacket to remove the stripped section. This is shown in Figure 3.3.5.

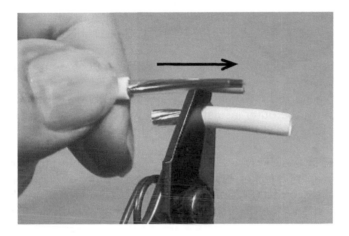

Figure 3.3.5   Removing the outer jacket.

Yikes! I moved too fast for Ken (my photographer) to catch me! What you see here is the aftermath of a series of actions. I moved the jaws of the strippers over to the right, so they're over an uncut part of the outer insulation jacket. By doing this, I avoid any risk of nicking or gouging the inner conductors.

Once the jaws were correctly positioned, I pulled on the section of outer jacket past the cut, to remove it (in the direction of the arrow). I got real lucky, and the inner foil shield came away cleanly at the same time (see the part in Section 1 on stripping wire).

Regrettably, despite my luck, I have to cut away more of the foil shield underneath the outer insulating jacket, to insure a clean connection.

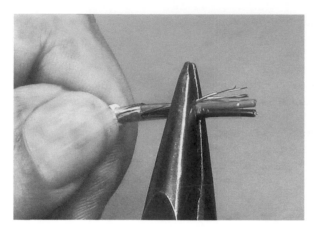

Figure 3.3.6   Pulling back the outer jacket.

Grasp the wire with one hand and gently pull equally on all the exposed conductors with the other (Figure 3.3.6). Here, I'm using a pair of needle-nose pliers that have smooth jaws to pull with, but your fingers will work too.

Our goal here is to expose more of the shield – maybe {1/8} inch – so that we can cut it away. By doing this we can use the outer jacket as an insulator and not have to deal with heat-shrinking anything.

Nick the foil, either with a razor blade or a pair of dykes. If you're not sure what I mean, look at Figure 3.4.10. I admit, I should have a shot here of the 'nick' – but I forgot – and the picture in the next section uses different colored wire, so I can't cheat and duplicate it.

Once you've started the tear, peel away the foil shield (Figure 3.3.7). When done properly, the outer insulating jacket will snap back to overlap the breakout of the individual conductors and the remaining foil.

Figure 3.3.8 shows the wire with the foil properly trimmed to recess under the outer insulating jacket. No foil fragments or other insulation are visible, so there's no chance of a short or of contaminating a solder point. I've also spiral-twisted the strands of the drain conductor, so they can be dealt with later.

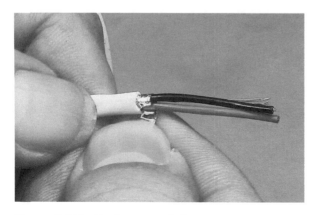

Figure 3.3.7   Peeling away foil.

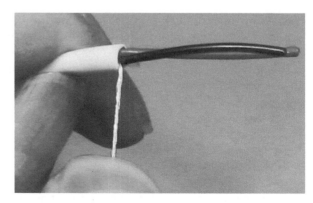

Figure 3.3.8   Completed basic strip.

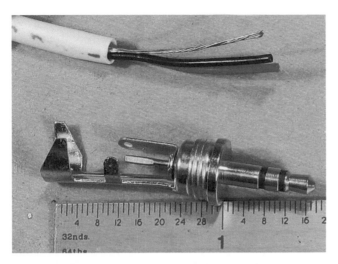

Figure 3.3.9   Wire against connector and ruler.

The wire is shown alongside the connector and a ruler in Figure 3.3.9.

The (roughly) 1 inch strip-back leaves oodles of slack to wire the MMGP. We'll wind up using less than {1/3} inch of the stripped wire when we're done, but the extra length makes it easier to work with.

Drat! I did it again – cut the darn conductors to length and forgot to take a shot of the actual cutting. Mea culpa. But you have a very vivid imagination, right? You can visualize the cruel, sharp jaws of the dykes shearing away the excess length, can't you?

The key thing to see here is that the high and low conductors have been cut to different lengths (Figure 3.3.10). The high (red) is about {9/32} inch and the low (black) is around {11/32} inch.

Each make of MMGP is a slightly different shape, so you'll have to guesstimate for the MMGP you're using. The point is that the different lengths need to match the dimensions of your particular MMGP, whatever they may be. Longer is better than shorter – you can always cut back.

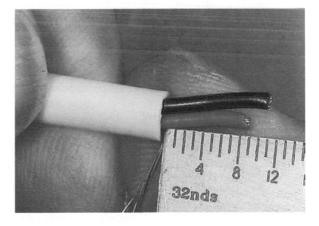

Figure 3.3.10   High and low conductors cut to fit.

Strip the ends of the high and low conductors with a pair of wire strippers (Figure 3.3.11). I gauge the distance by the width of the jaws of the hand stripper, which is about {3/32} inch. Yes, I know that's a tight tolerance, and no one will kill you if you err and make it {1/8} inch.

The reason to make such a short strip here is that we want to use a special soldering technique on the MMGP called 'beading'. In 'beading' we make a small blob (bead) of solder on the end of the conductor(s), and allow the conductor itself to carry the solder to the solder point (where the conductor and connector are joined). More on that later.

Figure 3.3.12 shows the completed strip on the high and low conductors. As you can see, I've nailed the {3/32} inch strip length pretty darn well. So you can too – it just takes a little practice and a little patience.

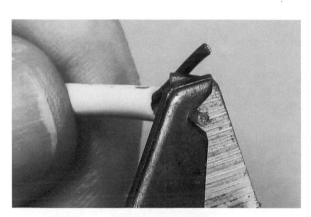

Figure 3.3.11   Stripping conductor ends.

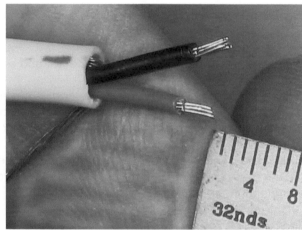

Figure 3.3.12   Stripped conductor ends.

Listen up, beading is a professional technique. It requires some delicacy, some practice, and *very* careful attention to safety details, if you don't want to get hurt.

Liquid solder flows rapidly and, in strict conformance with the Law of Gravity, it flows down – down onto your fingers, hand or leg if you're not careful. So pay attention to where your body parts are when beading conductors. Don't bead over your legs without a nice thick table in between. Do wear safety glasses and don't wear your Bermuda shorts to a wiring session.

Having warned you, please pay close attention to Figure 3.3.13 – it contains a lot of information about beading techniques.

Notice the orientation of both the wire and the individual conductors. Everything is pointing down toward the floor. Why? So our piping hot liquid solder will flow down toward the tip of the conductor. There, as it cools, it will form the round, blobby bead we want to create.

Any other orientation of the wire/conductors won't work. The molten solder will flow into unwanted places – maybe onto you – and your spoken vocabulary of rude phrases will be rapidly expanded.

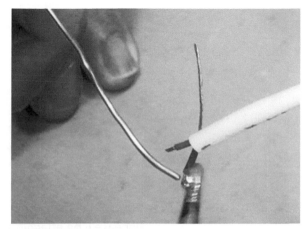

In Figure 3.3.13 I'm applying the soldering iron from underneath, but often the best approach is to have all the elements – the conductor, the solder and the iron – pointing down. Then the feeble-minded liquid solder has nowhere to go but down – down onto the tip of the conductor, where we want it to go. Not sure what I mean? Hold up the first three fingers (no thumb) of either hand toward the ceiling – like a mutant peace sign – and you'll get the concept.

Figure 3.3.13 Beading the conductors – 1.

It's also really important to pay attention to how much solder you're adding to the bead being formed. Too little solder and the bead will not have enough solder to bond properly. Too much solder and you will either create a bead that is too large or the solder will fall off in a wicked hot droplet that will burn a nice little crater in your leg. Sometimes, losing a few drops until you get a proper bead is the only thing that works. So always bead over that table I mentioned earlier.

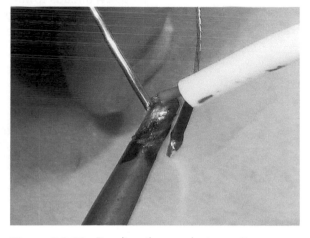

The same applies to beading the second conductor: some heat, just enough solder, and move the iron gently away from the conductor, so the molten solder will stay on the conductor and not transfer over to the iron tip (Figure 3.3.14). Now, on to the finished product.

Figure 3.3.14 Beading the conductors – 2.

Figure 3.3.15 shows the two conductors after beading. The black one has an optimal bead for our needs. The red one has a bead that's a little skinny, but still functional. Both beads are crusted over with flux – and this is (actually) a good thing. When reheated, the flux will help the solder on the conductor to meld with the tinning on the connector. Remember tinning?

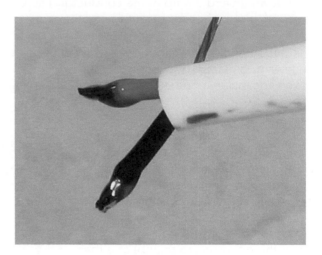

Figure 3.3.15   Completed beading.

Since we'll be attaching the wire to the connector next, *now* is the time to put the barrel over the wire (Figure 3.3.16). Make sure it's oriented with its screw threads toward the tip of the connector. Make sure to insert any insulating sleeves included with the connector. Since our subject connector didn't have a sleeve, we'll have to make one at the right time.

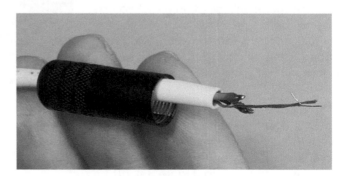

Figure 3.3.16   Putting barrel on wire.

Put the wire and barrel aside for now. We need to get delicately destructive. The particular MMGP I'm using has a 'solder tab' for the drain wire. But we don't need it, and it takes up space in our cramped micro-environment.

So I'm cutting it away with a strong pair of flush-cutting dykes (Figure 3.3.17). Don't try this with mini-dykes – you'll break the tool. Shear the tab (if there is one) away flush to the edge of the strain relief arm. No tab? No problem! Just ignore Figures 3.3.17 and 3.3.18.

Figure 3.3.17 Cutting off solder tab.

Figure 3.3.18 Tab sheared away.

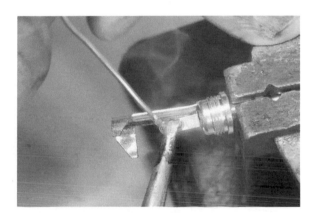

Figure 3.3.19 Tinning low solder tab.

Wow! My shearing cut was so close to the strain relief arm that I don't even have any sharp edges to clean up. If you are less fortunate, use a small file to smooth off any rough points. We don't want anything that will puncture the insulation of the sleeve we're going to add – well, make, in this case.

Just as we beaded the conductors, we want to melt a careful amount of solder onto the solder tabs and strain relief arm. See Figures 3.3.19 through 3.3.23. Then when the conductor and the solder tab are heated, the solder on both will meld into a physically strong and electrically conductive solder joint.

If everything is done properly, this will happen quickly, with only 2–3 seconds of heating from the soldering iron. The correct amount of solder to add is a thick film, or a slightly rounded, curved amount. We don't want an actual bead here, just enough solder for the bead on the conductor to bond with.

A correctly tinned low solder tab, ready for connection, is shown in Figure 3.3.20. Note the slightly rounded appearance of the solder – enough, but not too much. This is the low tab, but we have to tin both tabs and the strain relief arm. Which brings us to our next group of shots.

Tinning the high solder tab is the same as for the low tab (Figure 3.3.21). A little heat, a little solder and, voila, a gently rounded tin job. Bright and shiny, ready to bond.

Figure 3.3.20   Correct low tab tinning.

Figure 3.3.21   Tinning high solder tab.

Figure 3.3.22   Correct high tab tinning.

I've used a little more solder here (Figure 3.3.22), giving a more rounded tin job than the one on the low tab. The reason for this is that the high tab is so small; we need more solder for physical strength. The low tab, being larger and broader, can have a thinner coating.

The third, and last, of our tinning operations is to tin the strain relief arm (Figure 3.3.23), so that we can solder the drain/ground conductor to it easily. I'm going to create a fairly broad tinned area, but with a thin coating of solder. That way, no matter (almost) where the drain conductor winds up being soldered, it will have a nice film of solder to bond with.

Figure 3.3.23   Tinning strain relief arm.

Figure 3.3.24 illustrates the strain relief arm without the soldering iron, solder, and puffs of rosin smoke obscuring your view. Note the slightly rounded, bright, shiny appearance – exactly what we want the tinning to look like.

Figure 3.3.24   Correct tinning of strain relief.

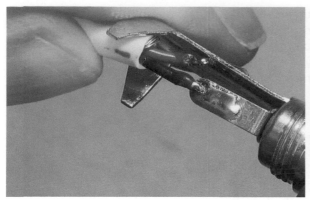

Figure 3.3.25 Positioning low conductor bead.

Figure 3.3.26 Soldering low conductor.

In Figure 3.3.25 I'm laying the bead at the end of the low conductor over the tinned end of the MMGP's low solder tab. I'll only get one shot at soldering it in place, without having to re-bead and/or re-tin, so correct physical orientation of the components is critical. Once everything is properly aligned, a quick touch of the soldering iron and the solder flows together from the bead onto the tab; the joint is complete. This is shown in Figures 3.3.26 and 3.3.27.

Press down firmly but gently on the solder bead. It will take (typically) no more than 2–3 seconds for the solder to heat up and meld between the conductor and the tab. While keeping the wire totally motionless, pull the iron quickly away and wait 5–10 seconds for the molten solder to cool. Any movement of the conductor or connector during this cooling period will degrade the connection.

See how nicely the low conductor and the solder tab have been joined? The solder has flowed into a smooth, shiny, slightly rounded dome that provides excellent conductivity and a durable, physically strong connection. Even better, once the first solder joint is made, the conductors and connector become self-supporting and self-aligning, making the following solder joints easier to create. Aren't you happy?

Figure 3.3.27 Finished low conductor joint.

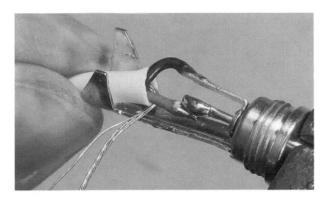

Figure 3.3.28   Positioning high conductor bead.

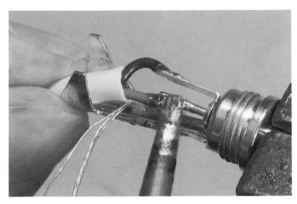

Figure 3.3.29   Soldering high conductor.

Just as we did with the low conductor bead, align the high conductor bead carefully over the high solder tab (Figure 3.3.28). Since the high tab is so small, we'll lay the iron across both the tab and bead, to heat both equally, as we see in Figure 3.3.29.

Just as described above, heat the high conductor bead and the high tab at the same time. The same instructions apply as those for soldering the low tab. Press down firmly but gently on the solder bead (and, in this case, on the tab as well). It will take (typically) no more then 2–3 seconds for the solder to heat up and meld between the conductor and the tab. While keeping the wire totally motionless, pull the iron quickly away and wait 5–10 seconds for the molten solder to cool. Any movement of the conductor or connector during this cooling period will degrade the connection.

Figure 3.3.30   Finished high conductor joint.

Again, you can see the well-formed shape of the finished solder joint (Figure 3.3.30). Rounded, smooth, shiny, and with enough solder to provide good physical strength. With both the high and low conductors in place, we have only to solder the drain conductor. But its strands have become untwisted and frazzled. We have to retwist them first.

I'm (re)twisting the drain conductor strands, and also putting the wire into position within the yoke of the strain relief arm (Figure 3.3.31). I want the outer insulation jacket to be as close as possible to the inner surface of the yoke, but not touching it. Why? Because the strain relief arm will become very hot when the drain conductor is soldered. If the insulation jacket is in contact with it, the insulation may be melted away. Not cool (literally).

We need to make a weight to hang off the drain strands, one that will keep them in position for soldering. Figure 3.3.32 shows a pair of pliers with a rubber band for some clamping action – easy to improvise.

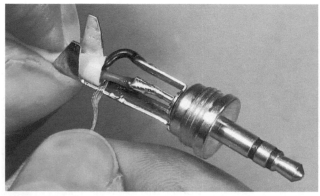 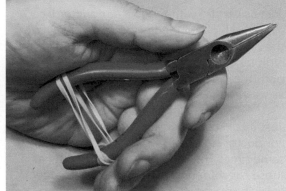

Figure 3.3.31 Twisting drain conductor strands.   **Figure 3.3.32** Create weight for drain.

You can use anything that has sufficient weight to pull down on the drain strands – hemostatic forceps, pliers or an old fishing sinker. Be creative – the point where the weight is attached will be cut off the drain strands after soldering anyway.

The truly perspicacious reader will note that I'm saying 'drain' here, and I used the term 'shield' when talking about the same operation for the full-size {1/4} inch guitar plug. Why? For the {1/4} inch plug, drain and shield were the same thing, because of the type of wire we were using. Here, we're using a different kind of wire – with that mylar shield, remember? So the mylar is the shield, and the strands of the drain conductor – which are in electrical contact with the mylar along its entire length – are the drain. I hope you don't find these minor distinctions too, ah, draining.

I've rotated the connector in the vise so the strain relief is facing upward. Next, I wrapped the drain strands tightly over the arm of the strain relief and clamped the jaws of the pliers on them.

Our weight here is the pair of pliers hanging off the drain strands. See how tightly the strands are wrapped across the strain relief arm in Figure 3.3.33?

Figure 3.3.33   Drain and weight in position.

Also note that the wire has fallen away from the yoke, which is a good thing – that way, the yoke won't melt the outer jacket when we solder. All is in position for the last solder point to be done.

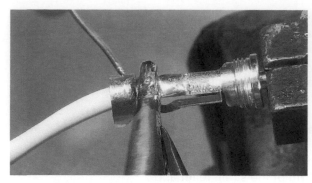

Figure 3.3.35   Correct soldering of drain.

Figure 3.3.34   Soldering drain strands.

Press firmly down with the tip of the iron for maximum heat transfer (Figure 3.3.34). Be liberal with the amount of solder you run in; here we want to fill all the voids in the drain strands, so they and the arm of the strain relief become one piece of metal. Just don't use so much solder as to cause blobs to appear on the drain strands.

See the rounding of the solder on the drain strands in Figure 3.3.35? There's enough solder to give us our desired electrical and physical properties, but not so much as to cause problems later. Now we have to remove the excess drain strands past the solder point.

Use dykes to cut away the excess drain strands (Figure 3.3.36). Cut as close as possible to the arm of the strain relief. Also cut at an angle, so the cut edge is beveled (angled) in relation to the strain relief arm. It gives you less to file down. I got lucky and my beveled cut gave me a nice, smooth edge. If you're less fortunate, file away, but do it carefully.

Figure 3.3.36   Removing excess drain strands.

Figure 3.3.38   Clamping yoke onto wire – 2.

Figure 3.3.37   Clamping yoke onto wire – 1.

Using any heavy duty pair of pliers, bend one arm of the strain relief yoke around the wire (Figure 3.3.37). Work slowly to form the arm into an arc. The goal is to immobilize the wire without crushing or deforming it, so just a slight amount of compression is the desired amount.

Figure 3.3.38 shows the desired curvature of the yoke arm and the optimal amount of compression. Once the first arm is formed correctly, we can do the same operation on the other arm – as shown in Figure 3.3.39.

Do the same action on the other yoke arm. Wrap it firmly but gently around the wire, letting it overlap (curl around) the first yoke arm. I'm showing this action is such detail because if you get overenthusiastic, and apply too much force, you'll crush the conductors, create a short and have to do the whole connector all over again. That's very boring, so be careful.

A perfectly formed yoke is illustrated in Figure 3.3.40, showing the right amount of compression and the correct curvature. It's not really that hard to do, and after you've done a half dozen connectors you'll wonder why you ever thought it was difficult.

Figure 3.3.39   Clamping yoke onto wire – 3.

Figure 3.3.40   Correctly formed yoke.

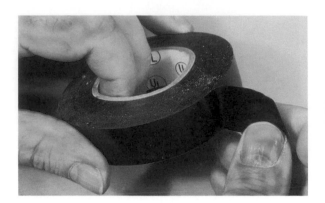

Figure 3.3.41   Electrical tape for insulation.

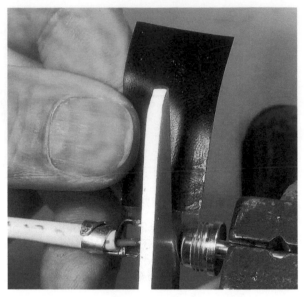

Figure 3.3.42   Cut tape width to fit MMGP.

Remember when we started working on this MMGP? I found there was no insulating sleeve included with this particular connector. But we need to create some kind of insulation between the (metal) outer barrel and the conductors/solder tabs inside. Electrical tape works nicely for this purpose (Figure 3.3.41).

In an emergency, you could use Scotch tape, masking tape, gaffer tape, even a band-aid, or paper wrapped around the solder tabs. But you must use something to avoid potential shorting of the tabs to the barrel. All of this is moot if you have an MMGP with a plastic barrel – but those break easily and don't provide the RF shielding of a metal barrel. So try and use metal barrel connectors – it's worth the extra work.

I'll use a couple of inches of electrical tape – it's the best of the alternatives at hand. To do so, I'll pull back about 2 inches of tape and cut it off the roll, for easy handling. Length is not critical here; anything longer than an inch and a half is fine.

Whoops, the width of the electrical tape is too wide to fit into the space in the MMGP. No problem, just cut away the excess (Figure 3.3.42); you can do this by eye, as the amount needed is not too critical.

Next we need to wrap tape around the solder tabs (Figure 3.3.43). This is a little tricky to explain, but pretty simple to actually do. I've poked the tape into the MMGP, so it passes between the high solder tab and the strain relief arm. Then I pulled the tape through and curled it around so a layer lies between the high and low solder tabs. If you can imagine the cross-section of the tape being shaped like the number '9', with the low tab inside the closed part of the '9' and the high tab resting in the open part of the '9'. I'll take the 'tail' (open part) of the '9' and wrap it around both the other layers and the strain relief arm itself.

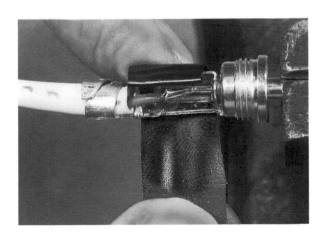

Figure 3.3.43 Wrap tape around solder tabs.

When this is all done, there will be at least one layer of tape between each of the solder tabs, and anything else, along with a couple of layers wrapped around the outside. No more is necessary and would actually hinder screwing the outer barrel into position.

Figure 3.3.44 Tape wrap completed.

The completed tape wrap is shown in Figure 3.3.44. There is enough to insulate, but not so much as to prevent the outer barrel being put in place. And isn't it nice that you now know a technique for insulating the solder tabs when the factory-supplied insulating sleeve is missing? Show this trick to someone assembling MMGPs and they might call you a star wireperson, maybe even a 'wrap' star!

Slide the outer barrel down and screw it home. With any luck, your completed MMGP connector will look very similar to the one in Figure 3.3.45, and work properly for a long, long time. But there's one final step left to ensure such longevity.

Figure 3.3.45 Completed MMGP.

Figure 3.3.46 shows the tip, ring and sleeve being sprayed with an electrical enhancer. I'm using Caig ProGold G5 and recommend that you do too! It prevents aerobic corrosion – which degrades conductivity. This is especially important when using miniature connectors, as the contact area is so small.

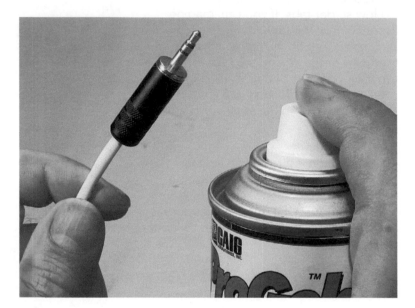

Figure 3.3.46   Spraying with ProGold.

ProGold improves electrical contact on all metals used in electrical connections, and is especially good for low-power connections. At low levels, any slight reduction in conductivity can be much more harmful than it would be at line level or speaker level. At higher power levels, the signal can cut through layers of (semi-conductive) corrosion.

Note: Caig Labs has recently changed the name from ProGold to DeoxIT Gold. It's the same stuff. They also have some new products that are interesting. More info about ProGold is on the Caig Labs website (www.caig.com).

That's it, folks. 'That's all she wrote' – or maybe I should say 'wired'. Now it's time to whip out that VOM you bought after I suggested you do so in Section 1. If you really read the whole book, you'll recall a mini-course in how to use it (the VOM) that was part of Section 2. You'll even remember what VOM, or DVOM, stands for.

Test your work, check it for shorts and high-resistance solder points. If it all checks out, enjoy your new connections!

3.4 RCA male connectors

The RCA connector was designed and popularized by (who else) the Radio Corporation of America – hence the name. Back in the 1940s and 1950s home hi-fi equipment was a new field, and there weren't too many standards or equipment makers.

So when RCA needed a small, low-cost connector for the equipment they were manufacturing, there wasn't anything currently on the market that quite fitted their needs. They designed a new plug and it became an industry standard.

RCA connectors are strictly for unbalanced wire runs, i.e. one-conductor shielded wire. You'll see them on stereo gear, home theater components, semi-pro and pro audio and video gear, and that dying class of equipment: turntables for vinyl records. You'll never see a balanced RCA connector – they don't exist. They have tip and shield only; never tip, ring and shield, like a stereo guitar plug. Hot and low/shield only, got it?

The signals entrusted to these humble connectors are as varied as their locations. An RCA may carry audio, video or digital information. It can do so at very low voltage and current – like the output of a turntable cartridge. Or it may carry medium-level signals, either audio or video. Some digital equipment also uses RCAs for input/output.

You'll seldom see RCAs used for speaker-level (high-level) connections, and for good reason. It's a wimpy connector, not designed to take high power levels or physical abuse. The main exception to this rule is in computer-type speakers, where the speakers themselves have a built-in amplifier. You may then see RCAs on the speaker(s) for the input from a computer's sound card, or another line-level device.

Now that I've thoroughly bad-mouthed these poor connectors, let's look at one in the flesh, or rather in the metal. Good quality RCAs have a metal outer barrel, *not* a plastic one! Don't use RCAs with a plastic outer barrel, except as a last resort; they don't provide the shielding of a metal barrel.

The RCA shown in Figure 3.4.1 is made by Canare; it's got a metal barrel and a built-in strain relief for the wire. Not all RCAs are so evolved; if you work on early gear, you may see much more primitive versions. Let's see how we get ready to wire this little puppy!

Figure 3.4.1   Male RCA.

Wrap the wire firmly around your hand (Figure 3.4.2). This prevents the individual conductors from being pulled out when they are stripped, and is only needed on short lengths of wire. If the wire is already harnessed or strain-reliefed, you don't have to wrap it around your hand.

Measure the wire a convenient distance back. Figure 3.4.3 shows the wire measured against a single-edge razor blade, which gives a length of {3/4} inch, but any repeatable length up to about 1 inch will do. It is easier to dress the wire correctly if you strip it longer than needed and then cut off the excess.

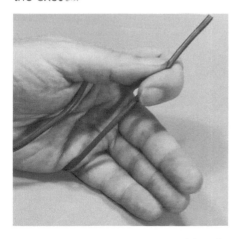

Figure 3.4.2   Wrap wire around hand.   **Figure 3.4.3**   Measure for strip-off.

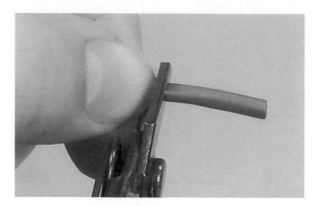

Figure 3.4.4   Cutting outer jacket with strippers.

Note that by placing my thumbnail at the point being measured, I can guide my stripping tool safely to the exact spot I want. This makes it easier to strip back an accurate distance each time.

The first part of the strip-off procedure is to cut into the outer insulation jacket enough to score it deeply, but not so hard as to nick the inner conductors (Figure 3.4.4). This can be done by 'feel' after you've done a few dozen.

Even better, set the guide on the strippers to the exact depth needed using scrap wire, before you start any real work. If you're using strippers, I'm going to ask you to move them over to a part of the outer jacket closer to the edge, before pulling on the outer jacket to remove the stripped section. This is shown in Figure 3.4.5.

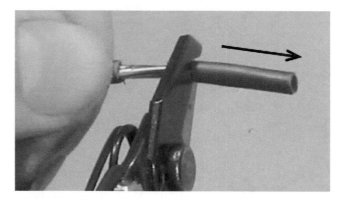

Figure 3.4.5   Removing outer jacket.

I've moved the jaws of the strippers over to the right, so they're over an uncut part of the outer insulation jacket. By doing this, I avoid any risk of nicking or gouging the inner conductors. Once the jaws are correctly positioned, pull on the section of outer jacket past the cut, to remove it. See the part in Section 1 on stripping wire.

Figure 3.4.6 shows the other way to strip the outer jacket – with a razor blade. OK, I admit it, you can't *see* the wire in this shot. But it's there, and the razor blade is touching the wire right at the point where the center of my left thumbnail is resting on the outer jacket. I added the arrow to inspire your imagination as to the placement of the wire.

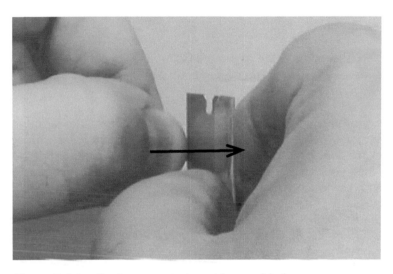

Figure 3.4.6   Cutting outer jacket with razor blade.

Strip the outer jacket off the wire with a pair of wire strippers (Figures 3.4.4 and 3.4.5) or use a razor blade (Figure 3.4.6) and gently cut around the insulation in a circular motion. If you've made the cut properly with a razor blade, you can pull the cut piece of outer jacket insulation off with your fingers.

When you are holding the wire correctly, your thumbnail will act as a safe and reliable guide, for the stripper or razor blade to meet the wire. I'm right-handed, so I use my right hand for the tool and left hand for the work (wire). If you are left-handed, just use your left hand for the tool and your right hand to hold the work.

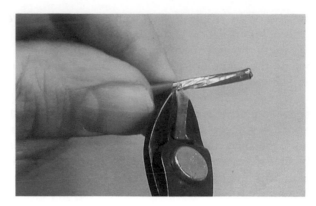

Figure 3.4.7   Nicking the foil.

Figure 3.4.7 shows the edge of the foil being nicked with a pair of dykes, so it can be removed cleanly. The same action can be done by slicing the foil gently with a razor blade.

It's helpful to grasp the blue mylar foil in one hand, the outer jacket in the other hand, and pull down the outer jacket so that more foil is exposed. Then nick the foil about {1/8} to {1/4} inch past the cut-off point of the outer jacket. When you release the outer jacket it will snap back, overlapping the foil cut and hiding any rough edges.

Pull on the foil where you nicked it and it will come away cleanly, exposing the wires (Figure 3.4.8). Clean up any rough edges by cutting them with dykes, and gently stretch the outer jacket back over the foil cut-off.

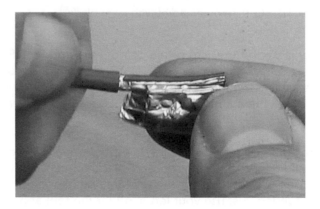

Figure 3.4.8   Removing the foil.

Pull on all three conductors to expose more shield (Figure 3.4.9). This is the first step in preparing a balanced wire for use with an unbalanced connector. An RCA connector has only high (often red) and shield/drain; there is *no* low (often black) conductor. Other color combinations are also frequently used.

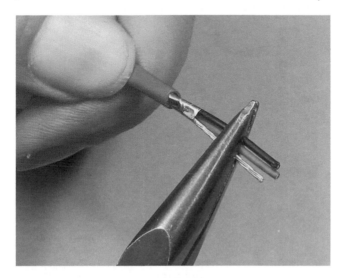

Figure 3.4.9 Expose more shield.

The question arises what to do with the unused low (black) conductor. For various esoteric reasons, I recommend ruthlessly cutting back *the* low conductor, *not* 'doubling up' low and drain.

The only exception to this is when the drain is only connected at *one* end. In that case, black and drain are connected at one end, but *only* black is connected at the other end. The drain then serves as an electrostatic shield, but not as a low-side conductor.

In most cases, RCAs are wired with one-conductor shielded (unbalanced) wire, so there is no low conductor to deal with. I show this method of using balanced wire on the assumption that it's easier for you to forget something than it is for you to remember it. Got balanced wire? Do what you see here. Got unbalanced wire? Take that low conductor and forget about it!

Once you are familiar with the construction of RCA connectors, and the specific type you're working with, the steps shown in Figures 3.4.8, 3.4.9 and 3.4.10 may be combined into one operation. I'm asking you to start by doing the work in discrete stages, so you can learn to do it accurately.

Nick the foil as before, just a little further back on the wire (Figure 3.4.10). Then tear the foil off where you nicked it. And no, I'm not going to show you the step of tearing off the foil again. I'm leaving that to your febrile imagination. Instead, I'll show you what to do after you've torn off the foil a second time and there are little nubbins of exposed foil left over.

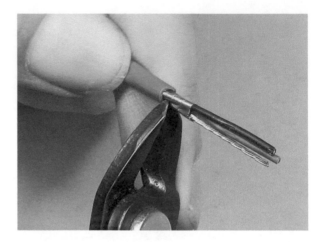

Figure 3.4.10   Nick foil again.

Carefully cut away those pesky nubbins of foil shield (if any) that are left after you've removed about {1/8} inch more of the foil (Figure 3.4.11). Be *very* careful not to cut, or nick, either the hot or drain conductor when you do this.

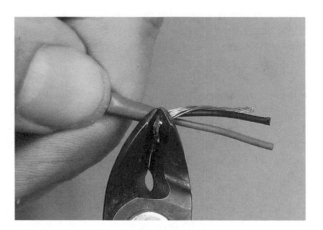

Figure 3.4.11   Remove foil scraps.

Cut off the low conductor as close to the breakout from the outer jacket as possible (Figure 3.4.12). Our goal is to have only the high and drain conductor exposed.

Figure 3.4.12   Remove low conductor.

Grasp the wire firmly in both hands. Stretch the outer jacket back toward the exposed conductors (Figure 3.4.13). This will create an overlap of the outer jacket, past the point where the foil and low conductor were cut away. By making this small insulating overlap, we reduce the risk of 'shorts' in the wire.

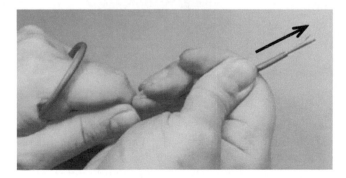

Figure 3.4.13   Stretch outer jacket back.

Figure 3.4.14 illustrates our spiffy dressed wire – no sign of foil or low conductor. Nice. We are now ready for the next steps.

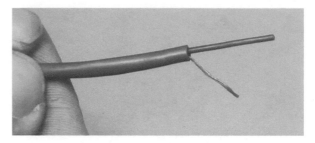

Figure 3.4.14   Ready for connection.

If you haven't wired many connectors, it's useful to hold the stripped wire against the connector to be soldered (Figure 3.4.15). This gives a hard-core reality check about how long the final length needs to be. When you're working from measurements, or have a clear understanding of the length needed, you can omit this step.

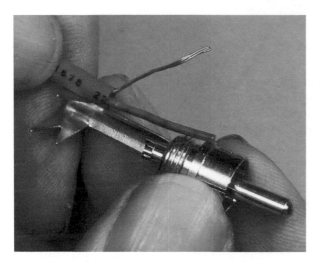

Figure 3.4.15   Length check by eye.

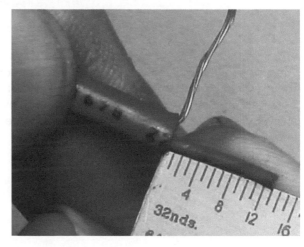

Figure 3.4.16   High conductor cut to length.

As can be seen in Figure 3.4.16, the high conductor has been cut to a tasteful {13/32} inch length. That's a really nice length for this particular Canare brand RCA, but there are literally dozens of different makers of RCAs. All these different models will have slightly different length requirements.

Use your eyes, use your rulers. Most of all, use your common sense. I can't possibly show all the variations – this book would be a bulky, unportable encyclopedia, and you'd likely drop it on your foot.

Strip the ends of the high conductor with a pair of wire strippers (Figure 3.4.17). I gauge the distance by the width of the jaws of the hand stripper, which is about {3/32} inch. If necessary, twist the strands of the hot and drain conductor (separately), so they will stay in place to be soldered.

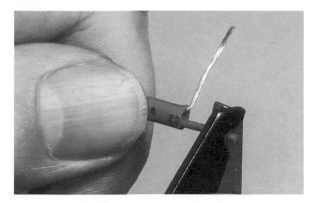

Figure 3.4.17   Stripping high conductor.

The finished strip-off is illustrated in Figure 3.4.18, showing {3/32} inch of exposed high conductor. This exposed length will increase when the conductor is tinned. The insulation melts back when heated – this is called 'wicking', or sometimes 'wickback'. Think of a candle's wick and you'll see the origin of the term.

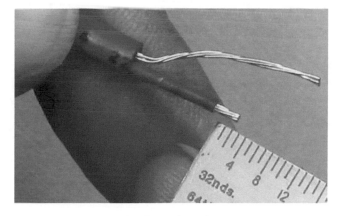

Figure 3.4.18   Finished strip length.

Speaking of tinning, wicking and suchlike, that's exactly what we're doing here. Tin *only* the high conductor (Figure 3.4.19). Do *not* tin the drain conductor – that would make it stiff and inflexible.

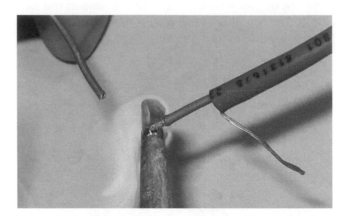

Figure 3.4.19   Tinning high conductor.

Putting a barrel on the wire is a *very* important step (Figure 3.4.20)! If you forget it, you'll have to *un*solder the wires from the connector, put the metal barrel onto the wire, and then *re*-solder the wires to the connector. Boring. So get it right the first time.

Figure 3.4.20   Putting barrel on wire.

Put the wire carefully aside. Mount the connector in a vise and tin the high (tip) solder cup (Figure 3.4.21). Fill it (almost) full with happy, molten solder, but don't get too enthusiastic – a gently brimming spoonful, right?

Figure 3.4.21   Tinning connector – 1.

The tinned high cup is shown in Figure 3.4.22, without the iron's tip obscuring the view. There is just enough solder. It's bright, shiny and smooth. Nice.

Figure 3.4.22   Tinning connector – 2.

Rotate the connector 180 degrees, so the back (yoke) is facing you. Tin the middle of the yoke (Figure 3.4.23). Slide the iron's tip along and keep adding dabs of solder until most of the yoke is nicely tinned. This allows the drain wire to be wrapped in a variety of positions and still lie on a tinned surface.

Figure 3.4.23   Tinning connector – 3.

Hurray! We finally get to attach the conductors (Figure 3.4.24). Heat the high solder cup on its top; heating will be too slow if you try from the bottom. When the solder is molten, quickly (but gently) slide the wire into the cup. Then gently slide the iron horizontally off – in this case, to the right – while holding the wire in place until the solder cools.

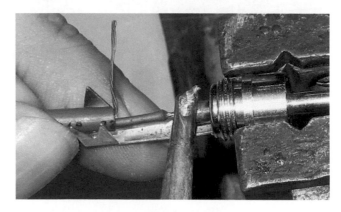

Figure 3.4.24   Attaching high conductor.

The completed high conductor is shown in Figure 3.4.25. Note that there are *no* exposed strands on the high conductor; the insulation is flush to the solder cup. Our careful measurement has resulted in a bit of the outer jacket protruding past the strain relief at the leftmost side of the connector. This is a good thing – when the strain relief is crushed down, the protruding outer jacket will add strength. Just wait and see.

Figure 3.4.25   Completed high conductor.

When you create a way to hold something in position that you're working on, that holding-in-place device is called a 'jig'. It's a way to position the work that makes dealing with it faster.

Here, by adding a rubber band to my pliers, we make a clamping action tool, one that's heavy enough to weigh down the drain conductor and keep it in position to be soldered (Figure 3.4.26). Thus, the vise holding the connector, and the pliers pulling down on the drain conductor, become a simple but effective wiring jig. This becomes very clear in Figure 3.4.27.

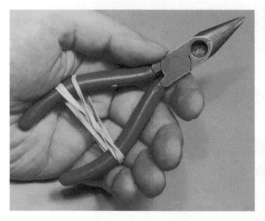

Figure 3.4.26   Pliers as a jig weight.

Figure 3.4.27   Soldering drain conductor – 1.

If you've ever heard the expression 'Done in jig time', it doesn't mean folks were dancing to an Irish tune. Rather, that they were clever enough to do their work with a jig, get it done quickly and have time left over.

Pay close attention now. A lot has happened up to the time the photo in Figure 3.4.27 was taken.

I rotated the connector around again, so the yoke is facing up. Then I wrapped the drain conductor up, around the yoke, and clamped the pliers' jaws on the drain conductor, past where it wrapped over the yoke. The weight of the pliers pulls down on the drain wire, holding it in position to be soldered. Be sure the outer jacket lies close to the other side of the yoke (the side *not* being soldered). With the drain in position, I soldered it in place with a gently generous dollop of solder and my trusty iron.

Figure 3.4.28 provides another view of the soldered drain conductor – flat, smooth and shiny, with enough solder to add physical strength. But what to do about that bit of leftover drain sticking off the side?

Figure 3.4.28   Soldering drain conductor – 2.

Another view of the soldered connector is shown in Figure 3.4.29. That pesky extra drain is still hanging around, but not for long. The wire lies flush inside the strain relief and there are no contaminating scraps of materials present.

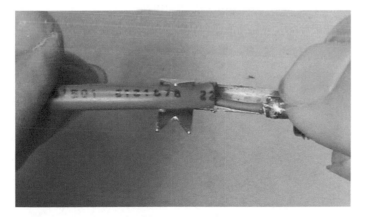

Figure 3.4.29   Soldering drain conductor – 3.

Carefully cut away the excess drain conductor (Figure 3.4.30). The best way to do this is to cut at an angle, so the cut edge is beveled, with no bump or sharp sides.

What's that? You left some sharp edges when you cut off the drain? No problem, just file them flat and smooth (Figure 3.4.31). Here I'm using the file from my Swiss Army knife, but any file that fits will do.

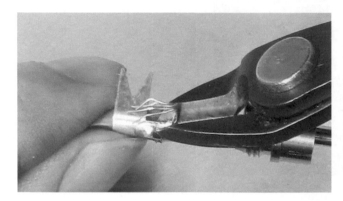

Figure 3.4.30   Cutting excess drain.

Figure 3.4.31   Filing drain conductor.

Figure 3.4.32 shows what you will (hopefully) wind up with after all this soldering, cutting and filing; i.e. no sharp edges and ready for the next step. You get to crush something (but very gently).

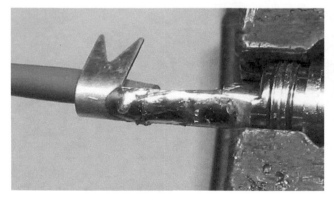

Figure 3.4.32   Filed drain conductor.

Figure 3.4.33   Crimping strain relief.

It may sound like an oxymoron, but it's not. (Figure 3.4.33) Gently crush the tabs of the strain relief around the wire. Do them one at a time, and the goal is to curl the metal tab around the wire. You could just pinch the tabs together, but that makes a weak strain relief which breaks soon. Curl, baby, curl!

Darn it, I forgot to show you how to screw the barrel onto the connector itself. But you can figure that out, right? It's just so automatic for me that I breezed right through it. Let's see, the threads on the inside of the barrel mate with the threads on the outside of the connector, if I turn clockwise.

In Figure 3.4.34 I've screwed the barrel on, and the RCA is shown being sprayed with enhancer. Spray the tip and inside the ring. Wipe off any excess.

I'm using Caig ProGold G5 and *you* should too! It improves electrical contact on all metals used in electrical connections, and is especially good for low-power connections.

Note: Caig Labs has recently changed the name from ProGold to DeoxIT Gold. It's the same stuff. They also have some new products that are interesting. More info about ProGold is on the Caig Labs website (www.caig.com).

Figure 3.4.34 Spray the RCA with enhancer.

The signal level on an RCA can range from medium strong to very weak, so any help you can give those hard-working electrons will be greatly appreciated!

The completed RCA male, in all its glory, is shown in Figure 3.4.35.

Now it's time to whip out that VOM you bought after I told you to do so in Section 1. If you really read the whole book, you'll recall a mini-course in how to use it (the VOM) that was part of Section 2. You'll even remember what VOM, or DVOM, stands for.

Figure 3.4.35 Completed RCA male connector.

Test your work, check it for shorts and high-resistance solder points. If it all checks out, enjoy your new connections!

RCA female connectors

The RCA female is very similar to the male connector, but there are enough differences to merit a separate section for it.

The RCA connector was designed and popularized by (who else) the Radio Corporation of America – hence the name. Back in the 1940s and 1950s home hi-fi equipment was a new field, and there weren't too many standards or equipment makers.

So when RCA needed a small, low-cost connector for the equipment they were manufacturing, there wasn't anything currently on the market that quite fitted their needs. They designed a new plug and it became an industry standard.

RCA connectors are strictly for unbalanced wire runs, i.e. one-conductor shielded wire. You'll see them on stereo gear, home theater components, semi-pro and pro audio and video gear, and that dying class of equipment: turntables for vinyl records. You'll never see a balanced RCA connector – they don't exist. They have tip and shield only, never tip, ring and shield, like a stereo guitar plug. Hot and low/shield only, got it?

The signals entrusted to these humble connectors are as varied as their locations. An RCA may carry audio, video or digital information. It can do so at very low voltage and current – like the output of a turntable cartridge. Or it may carry medium-level signals, either audio or video. Some digital equipment also uses RCAs for input/output.

You'll seldom see RCAs used for speaker-level (high-level) connections, and for good reason. It's a wimpy connector, not designed to take high power levels or physical abuse. The main exception to this rule is in computer-type speakers, where the speakers themselves have a built-in amplifier. You may then see RCAs on the speaker(s) for the input from a computer's sound card, or another line-level devise.

Now that I've thoroughly bad-mouthed these poor connectors, let's look at one in the flesh, or rather in the metal. Good quality RCAs have a metal outer barrel, *not* a plastic one! Don't use RCAs with a plastic outer barrel, except as a last resort; they don't provide the shielding of a metal barrel.

A female RCA with the shell in place is shown in Figure 3.5.1. What lies beneath the shell? Basically, a very similar structure to the RCA males of more recent design. There's a strain relief yoke, which also serves as a solder point, and a solder cup for the center (hot) conductor (Figure 3.5.2). Let's look at that now.

Figure 3.5.1   Female RCA – 1.

Figure 3.5.2   Female RCA – 2.

Notice something peculiar about the center solder cup? If you don't, flip back and look at the male of the species. That's right, the solder cup on the female is upside down, in relation to the solder cup on the male. And the male center cup is in the correct orientation. This is a *very* common design flaw, and most makers of RCAs will still make females this way. Some enlightened manufacturers have it right, but most don't. So I'm going to show you how to solder an upside-down center cup. Don't worry, it's not (too) hard.

Figure 3.5.3 gives a better view of the yoke, which is more or less the same as the male. I've included the butt end of a ruler in the picture to give a reality check of the size.

Figure 3.5.3   Female RCA – 3.

Wrap the wire firmly around your hand (Figure 3.5.4). This prevents the individual conductors from being pulled out when they are stripped, and is only needed on short lengths of wire. If the wire is already harnessed or strain-reliefed, you don't have to wrap it around your hand.

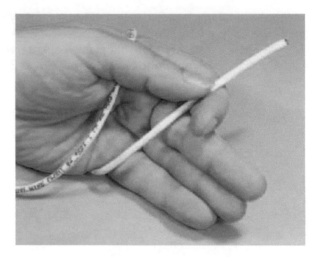

Figure 3.5.4   Wrap wire around hand.

Measure the wire a convenient distance back. Figure 3.5.5 shows the wire measured against a single-edge razor blade, which gives a length of {3/4} inch, but any repeatable length up to about 1 inch will do. It is easier to dress the wire correctly if you strip it longer than needed and then cut off the excess.

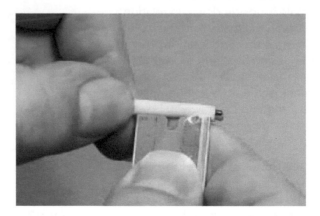

Figure 3.5.5   Measure for strip-off.

Note that by placing my thumbnail at the point being measured, I can guide my stripping tool safely to the exact spot I want. This makes it easier to strip back an accurate distance each time.

The first part of the strip-off procedure is to cut into the outer insulation jacket enough to score it deeply, but not so hard as to nick the inner conductors (Figure 3.5.6). This can be done by 'feel' after you've done a few dozen.

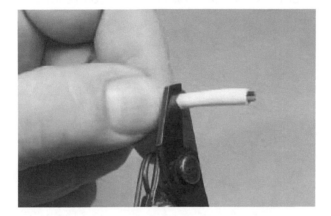

Figure 3.5.6   Cutting outer jacket with strippers.

Even better, set the guide on the strippers to the exact depth needed using scrap wire, before you start any real work. If you're using strippers, I'm going to ask you to move them over to a part of the outer jacket closer to the edge, before pulling on the outer jacket to remove the stripped section. This is shown in Figure 3.5.7.

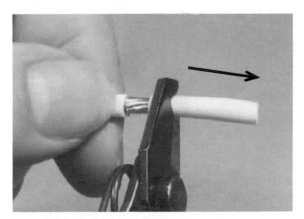

Figure 3.5.7   Removing outer jacket – 1.

I've moved the jaws of the strippers over to the right, so they're over an uncut part of the outer insulation jacket. By doing this, I avoid any risk of nicking or gouging the inner conductors. Once the jaws are correctly positioned, pull on the section of outer jacket past the cut, to remove it. See the part in Section 1 on stripping wire.

I was lucky with this strip – the foil came away cleanly (Figure 3.5.8). Even if you are not so fortunate, it's OK as we have a second shot at the foil shortly.

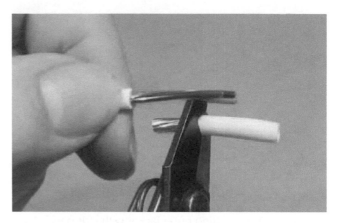

Figure 3.5.8   Removing outer jacket – 2.

I want to expose a bit more foil so I can clean it away, and do this without damaging the conductors of the wire. So I'm gently pulling on the conductors, while holding the outer jacket firmly in my left hand (Figure 3.5.9).

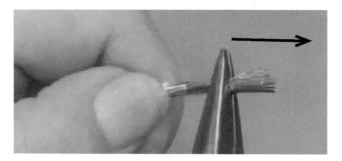

Figure 3.5.9   Pull on strands.

Pull on all three conductors to expose more shield. This is the first step in preparing a balanced wire for use with an unbalanced connector. An RCA connector has only high (often red) and shield/drain; there is *no* low (often black) conductor. Other color combinations are also frequently used.

The question arises what to do with the unused low (black) conductor. For various esoteric reasons, I recommend ruthlessly cutting back the low conductor, *not* 'doubling up' low and drain.

The only exception to this is when the drain is only connected at *one* end. In that case, black and drain are connected at one end, but *only* black is connected at the other end. Drain then serves as an electrostatic shield.

In many cases, RCAs are wired with one-conductor shielded (unbalanced) wire, so there is no low conductor to deal with. I show this method of using balanced wire on the assumption that it's easier for you to forget something than it is for you to remember it. Got balanced wire? Do what you see here. Got unbalanced wire? Take that low conductor and forget about it!

I show the removal of the black (low) conductor a bit later on.

Figure 3.5.10 shows the edge of the foil being nicked with a pair of dykes, so it can be removed cleanly. The same action can be done by slicing the foil gently with a razor blade.

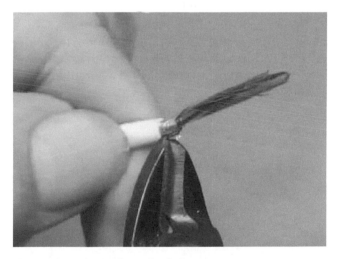

Figure 3.5.10   Nicking the foil.

As we showed in Figure 3.5.9, it's helpful to grasp the blue mylar foil in one hand, the outer jacket in the other, and pull down the outer jacket so that more foil is exposed. Then nick the foil about {1/8} to {1/4} inch past the cut-off point of the outer jacket. When you release the outer jacket it will snap back, overlapping the foil cut and hiding any rough edges.

Pull on the foil where you nicked it, and it will come away cleanly, exposing the wires (Figure 3.5.11). Clean up any rough edges by cutting them with dykes, and gently stretch the outer jacket back over the foil cut-off.

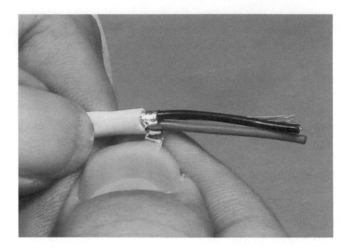

Figure 3.5.11   Removing the foil.

Twist the drain strands together so they don't get frazzled (Figure 3.5.12). We're not going to tin them, so we need to keep them together some other way. Tinning would make them stiff, and we want them to wrap tightly around the yoke.

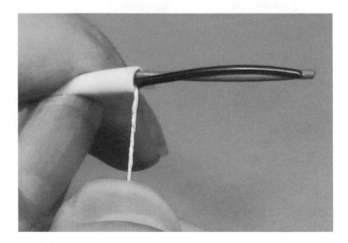

Figure 3.5.12   Twist drain strands.

Cut off the low conductor, as close to the breakout from the outer jacket as possible (Figure 3.5.13). Our goal is to have only the high and drain conductor exposed.

Figure 3.5.13   Remove low conductor.

Grasp the wire firmly in both hands. Stretch the outer jacket back toward the exposed conductors (Figure 3.5.14). This will create an overlap of the outer jacket, past the point where the foil and low conductor were cut away. By making this small insulating overlap, we reduce the risk of 'shorts' in the wire.

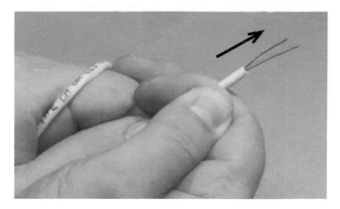

Figure 3.5.14   Stretch outer jacket back.

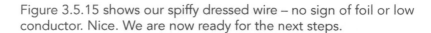

Figure 3.5.15 shows our spiffy dressed wire – no sign of foil or low conductor. Nice. We are now ready for the next steps.

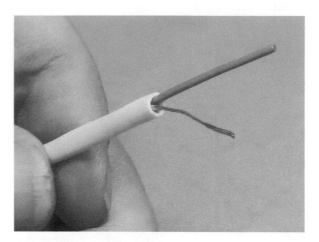

Figure 3.5.15   Ready for connection.

Putting the barrel on the wire is a *very* important step (Figure 3.5.16)! If you forget it, you'll have to *un*solder the wires from the connector, put the metal barrel onto the wire, and then *re*-solder the wires to the connector. Boring. So get it right the first time. OK, I admit I showed this step differently for the male RCA. The key is to do it before soldering.

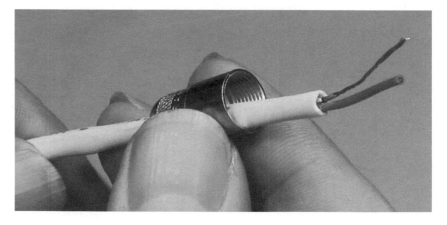

Figure 3.5.16   Putting barrel on wire.

If you haven't wired many connectors, it's useful to hold the stripped wire against the connector to be soldered (Figure 3.5.17). This gives a hard-core reality check about how long the final length needs to be. When you're working from measurements, or have a clear understanding of the length needed, you can omit this step.

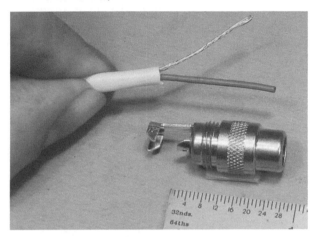

Figure 3.5.17   Length check by eye.

The length of the high (red) conductor is different here than the length used for the male RCA (Figure 3.5.18). The reason is that the two plugs have different dimensions, and the wire has to match the connector you're using, not a conceptual ideal. As you can see, the right length for this particular female RCA is {5/16} inch.

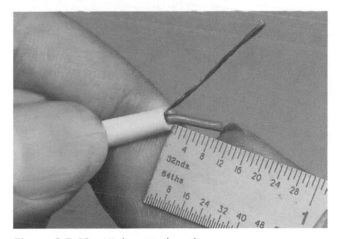

Figure 3.5.18   High cut to length.

Use your eyes, use your rulers. Most of all, use your common sense. I can't possibly show all the variations – this book would be a bulky, unportable encyclopedia, and you'd likely drop it on your foot.

And no, I didn't show you cutting the red conductor to {5/16} inch. I did it so fast Ken didn't have time to catch it. Force of habit – shoot me. But you can imagine it, right? Holding the red conductor with your thumbnail at the {5/16} inch point and sliding the steely jaws of the cutters down onto the red conductor? See, I knew you could do it!

Darn, I did it again – stripped {3/32} inch of insulation off the red conductor before Ken could blink or shoot (Figure 3.5.19). And that length of strip is important, because the exposed length will increase when the conductor is tinned. The insulation melts back when heated – this is called 'wicking', or sometimes 'wickback'. Think of a candle's wick and you'll see the origin of the term.

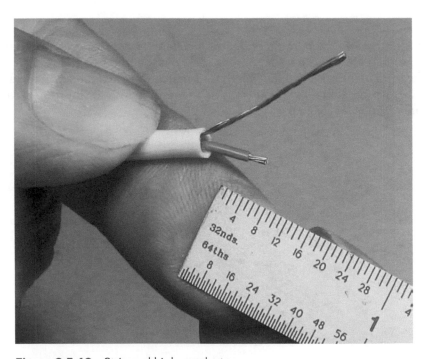

Figure 3.5.19   Stripped high conductor.

Speaking of tinning, wicking and suchlike, that's exactly what we're doing here. Tin *only* the high conductor (Figure 3.5.20). Do *not* tin the drain conductor – that would make it stiff and inflexible.

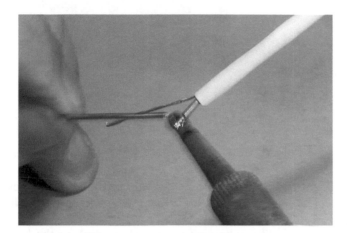

Figure 3.5.20   Tinning high conductor.

Put the wire carefully aside. Mount the connector in a vise, and tin the high (tip) solder cup (Figure 3.5.21). Fill it (almost) full with happy, molten solder, but don't get too enthusiastic – a gently brimming spoonful, right?

Figure 3.5.21   Tinning connector – 1.

Notice that I've inverted the connector? I want the molten solder to flow into the upside-down solder cup, so I flipped the connector. Always try to work *with* gravity, not against it – you'll be a lot less frustrated.

Figure 3.5.22   Tinning connector – 2.

Figure 3.5.23   Tinning connector – 3.

Figure 3.5.22 shows the tinned high cup, without the iron's tip obscuring the view. There is just enough solder. It's bright, shiny and smooth. Nice.

Rotate the connector (if you haven't already done this), so the back (yoke) is facing you. Tin the middle of the yoke (Figure 3.5.23). Slide the iron's tip along, and keep adding dabs of solder until most of the yoke is nicely tinned. This allows the drain wire to be wrapped in a variety of positions and still lie on a tinned surface.

Place the high conductor in position against the solder you put in the high conductor cup (Figure 3.5.24). Brace your fingers to keep it stable, and also keep your fingers away from the soldering iron that we're about to bring onstage.

Figure 3.5.24   Attaching high conductor – 1.

Hurray! We finally get to attach the conductors. Heat the high solder cup on its top; this connector has a thin wall on the center cup so we can do this (Figure 3.5.25). When the solder is molten, quickly (but gently) lift the wire into the cup. On connectors with a thicker center cup wall, you'll have to heat the tinned solder itself.

Figure 3.5.25   Attaching high conductor – 2.

Then gently slide the iron horizontally off – in this case to the right (away from my left hand), while holding the wire in place until the solder cools.

The completed high conductor is shown in Figure 3.5.26. Note that there are *no* exposed strands on the high conductor; the insulation is flush to the solder cup. Our careful measurement has resulted in a bit of extra length in the high conductor. This is deliberate – the extra length will be bent up in a hoop shape to provide strain relief. This is illustrated in Figure 3.5.27.

Figure 3.5.26   Completed high conductor.

Figure 3.5.27   Bend high conductor.

The high conductor is being bent into a hoop shape. The problem is, if I show the pliers well, you can't see much of the 'hoopishness' of it. What I'm doing is taking the wire's '−' shape and making more of an 'Ω' out of it. It's a good thing I had that omega symbol handy.

Our careful bending of the high conductor has resulted in a bit of the outer jacket protruding past the strain relief at the leftmost side of the connector (Figure 3.5.28). This is a good thing – when the strain relief is crushed down, the protruding outer jacket will add strength. Just wait and see.

When you create a way to hold something in position that you're working on, that holding-in-place devise is called a 'jig'. It's a way to position the work that makes dealing with it faster.

Figure 3.5.28   Bend complete.

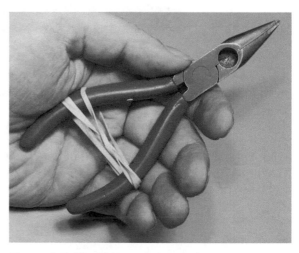

Figure 3.5.29   Pliers as a jig weight.

Here, by adding a rubber band to my pliers, we make a clamping action tool, one that's heavy enough to weigh down the drain conductor and keep it in position to be soldered (Figure 3.5.29). Thus, the vise holding the connector, and the pliers pulling down on the drain conductor, become a simple but effective wiring jig. This becomes very clear in the following figures.

If you've ever heard the expression 'Done in jig time', it doesn't mean folks were dancing to an Irish tune. Rather, that they were clever enough to do their work with a jig, get it done quickly, and have time left over.

Now I'm attaching the pliers to the drain conductor (Figure 3.5.30). Because of the rubber band, the pliers will have a strong but flexible clamping action and hold the drain conductor wrapped tightly around the yoke. Figure 3.5.31 shows this even more clearly. See how the pliers are hanging, suspended by the drain conductor? Exactly what we want to give a nice, tight wrap of the drain conductor around the yoke.

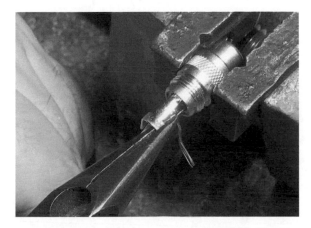

Figure 3.5.30   Grip drain with pliers – 1.

Figure 3.5.31   Grip drain with pliers – 2.

Pay close attention now. A lot has happened up to the time the photo in Figure 3.5.32 was taken.

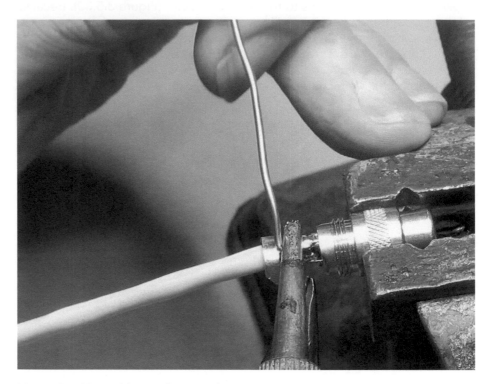

Figure 3.5.32   Soldering drain conductor.

I previously rotated the connector around again, so the yoke is facing up. Then I wrapped the drain conductor up, around the yoke, and clamped the pliers' jaws on the drain conductor, past where it wrapped over the yoke. The weight of the pliers pulls down on the drain wire, holding it in position to be soldered. Be sure the outer jacket lies close to the other side of the yoke (the side *not* being soldered). With the drain in position, I soldered it in place with a gently generous dollop of solder and my trusty iron.

Carefully cut away the extra drain conductor, flush to the edge of the 'arm' of the yoke (Figure 3.5.33). If needed, file the cut edge to smooth it down.

Figure 3.5.33 Cutting excess drain.

All is now ready for the final steps of assembly (Figure 3.5.34). The high and low/drain are soldered, and the outer jacket is far enough forward to have the strain relief wrap around it properly. Ready for some gentle crushing?

Figure 3.5.34 Ready for strain relief.

Gently curl the strain relief around the outer jacket and then crimp it in a bit (Figure 3.5.35) – just enough so the strain relief has a firm grip on the outer jacket, but not so much as to cut into it.

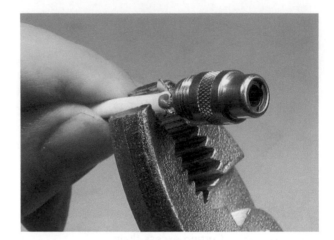

Figure 3.5.35   Curl strain relief.

The finished crimp is shown in Figure 3.5.36. See how the strain relief is tight enough to grip the outer jacket, without cutting it or crushing the conductors inside? We are now ready for the next step.

Figure 3.5.36   Finished crimp.

Rotate the barrel clockwise to screw it onto the body of the connector (Figure 3.5.37). But you knew that already, right?

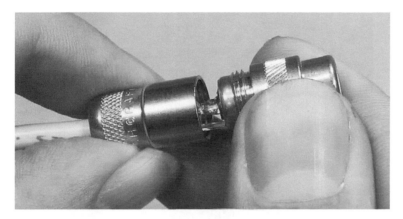

Figure 3.5.37 Screw on barrel.

The finished connector is shown in Figure 3.5.38: a thing of beauty and a joy forever – or at least until the contacts corrode. But we have a solution (literally) for that.

Figure 3.5.38 Finished connector.

In Figure 3.5.39, the RCA is shown being sprayed with electrical contact enhancer. Spray the tip and outside the ring. Wipe off any excess. I'm using Caig ProGold G5 and *you* should too! It improves electrical contact on all metals used in electrical connections, and is especially good for low-power connections.

Figure 3.5.39   Spray RCA with enhancer.

Note: Caig Labs has recently changed the name from ProGold to DeoxIT Gold. It's the same stuff. They also have some new products that are interesting. More info about ProGold is on the Caig Labs website (www.caig.com).

The signal level on an RCA can range from medium strong to very weak, so any help you can give those hard-working electrons will be greatly appreciated!

Now it's time to whip out that VOM you bought after I told you to do so in Section 1. If you really read the whole book, you'll recall a mini-course in how to use it (the VOM) that was part of Section 2. You'll even remember what VOM, or DVOM, stands for.

Test your work, check it for shorts and high-resistance solder points. If it all checks out, enjoy your new connections!

XLR female and male connectors

The XLR connector was originally designed for professional, high-level, balanced audio runs. It is also the most common connector for microphones and is sometimes called a 'microphone connector'.

With the advent of semi-pro gear, you can no longer trust an XLR connector to be balanced, high level or low impedance. You must determine, by looking at the equipment's manual and/or schematics, what the configuration of your equipment is. The pictures in this section show a three-wire balanced connection for an XLR three-pin female. Male connectors are also covered, but are a similar wire dress.

XLR connectors also come in other pin configurations. I have seen two-, three-, four-, five- and six-pin applications. However, the three-pin is by far the most common for audio and we will only consider that configuration here.

In the next sequence of pictures, I will show you a fast and reliable way to wire female and male XLR connectors.

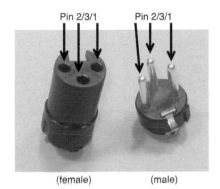

Pin 2/3/1 (female)　　Pin 2/3/1 (male)

Figure 3.6.1 Contact side of XLR.

Let's look at both sides of the connectors. To avoid confusion, I'll define the two sides as the 'contact' side and the 'wire' side. The contact side is the side that actually makes electrical contact with another connector. The wire side is the side we attach the wires to by soldering.

The instructions for the male XLR are the same as the instructions for the female, except that the placements of Pins 1 and 2 are inverted. The *wire* side of the female and male are mirror images of each other, like your two hands. Figures 3.6.1 and 3.6.2 will help to show what I'm talking about.

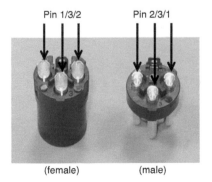

Pin 1/3/2 (female)　　Pin 2/3/1 (male)

Figure 3.6.2 Wire side of XLR.

At first, this seems a little confusing, but if you think about it, Pins 1 and 2 *have* to invert position, so that Pin 1 of the male will contact Pin 1 of the female when they are plugged together, as in Figure 3.6.3.

Well, they don't call them mating connectors for nothing. Normally the insertion of the male into the female is modestly cloaked by the outer shell.

Figure 3.6.3   Mated female and male.

Figure 3.6.4 shows both the male and female after wiring. The inversion of the hot (red) and shield (silver) conductor is easy to see. If you keep this simple flip on Pins 1 and 2 in mind, you'll never mis-wire an XLR.

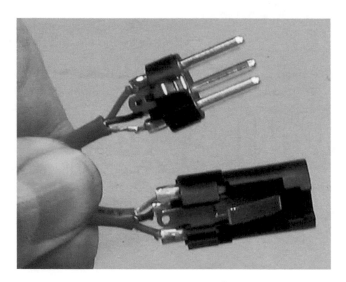

Figure 3.6.4   Wired female and male.

To make things even easier, both the male and female XLR have tiny little hard-to-read numbers next to the pins. So if you forget which is which, just grab a magnifying glass and really *look* at the plug you're soldering.

If you are breaking out multi-pair wire to connect several XLRs, please read the instructions in Section 1, starting on page 51, about removing the multi-pair outer jacket.

The instructions given here are for putting an XLR female on one wire, using wire that has two insulated conductors, one uninsulated drain conductor and a mylar foil shield – in other words, the most common wire found in studios. And don't worry folks, I also show pictures of the male XLR when needed to remind you of that pesky Pin 1/Pin 2 inversion.

The operations are similar for other types of wire, but it would be instructive for you to read the part in Section 1 on different types of wire if you are working with some of them.

In fact, if you have not read Sections 1 and 2, I suggest you do so before proceeding. This book is written in a logical, sequential order and the information in the first two sections lays the groundwork for the later wiring techniques demonstrated here.

The XLR plugs shown here are a Neutrik model. I recommend Neutrik over all other brands of XLR connectors, because of superior design features.

Further, I recommend *gold-plated* contacts on connectors over all other metals, no matter what brand you use. So a gold-plated ITT XLR is better than a nickel-plated Neutrik. Now a gold-plated Neutrik – that's really cool!

I don't own Neutrik stock nor any share in a gold mine. When I state a preference, it's because I think something is better, higher fidelity or more durable.

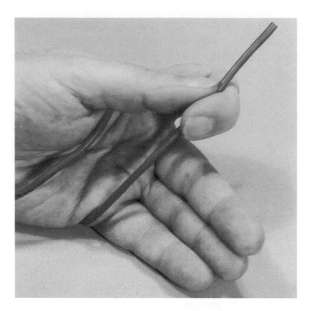

Figure 3.6.5   Wrap wire around hand.

Wrap the wire firmly around your hand (Figure 3.6.5). This prevents the individual conductors from being pulled out when they are stripped, and is only needed on short lengths of wire. If the wire is already harnessed or strain-reliefed, you don't have to wrap it around your hand.

Measure the wire a convenient distance back. Figure 3.6.6 shows the wire measured against a single-edge razor blade, which gives a length of {3/4} inch, but any repeatable length up to about 1 inch will do. It is easier to dress the wire correctly if you strip it longer than needed and then cut off the excess.

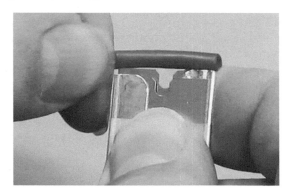

Figure 3.6.6   Measure for strip-off.

Note that by placing my thumbnail at the point being measured, I can guide my stripping tool safely to the exact spot I want. This makes it easier to strip back an accurate distance each time.

Cutting the outer jacket with strippers is the first part of the strip-off procedure (Figure 3.6.7). Cut into the outer insulation jacket enough to score it deeply, but not so hard as to nick the inner conductors. This can be done by 'feel' after you've done a few dozen.

Figure 3.6.7 Cutting outer jacket with strippers.

Even better, set the guide on the strippers to the exact depth needed using scrap wire, before you start any real work. If you're using strippers, I'm going to ask you to move them over to a part of the outer jacket closer to the edge, before pulling on the outer jacket to remove the stripped section. This is shown in Figure 3.6.8.

I've moved the jaws of the strippers over to the right, so they're over an uncut part of the outer insulation jacket. By doing this, I avoid any risk of nicking or gouging the inner conductors. Once the jaws are correctly positioned, pull on the section of outer jacket past the cut, to remove it. See the part of Section 1 on stripping wire.

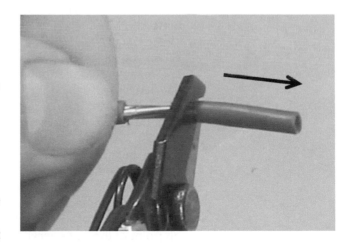

Figure 3.6.8 Removing outer jacket.

Figure 3.6.9 shows the other way to strip the outer jacket – with a razor blade. OK, I admit it, you can't *see* the wire in this figure. But it's there, and the razor blade is touching the wire right at the point where the center of my left thumbnail is resting on the outer jacket. I added the arrow to inspire your imagination as to the placement of the wire.

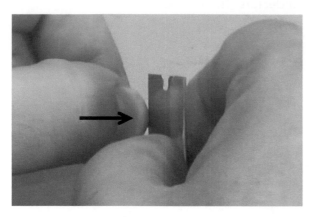

Figure 3.6.9   Cutting outer jacket with razor blade.

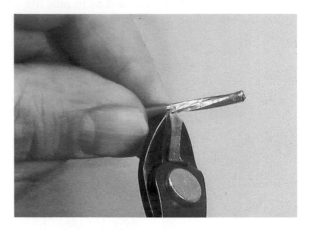

Figure 3.6.10   Nicking the foil.

Strip the outer jacket off the wire with a pair of wire strippers (Figures 3.6.7 and 3.6.8) or use a razor blade (Figure 3.6.9) and gently cut around the insulation in a circular motion. If you've made the cut properly with a razor blade, you can pull the cut piece of outer jacket insulation off with your fingers.

When you are holding the wire correctly, your thumbnail will act as a safe and reliable guide for the stripper or razor blade to meet the wire. I'm right-handed, so I use my right hand for the tool and my left hand for the work (wire). If you are left-handed, just use your left hand for the tool and your right hand to hold the work.

Figure 3.6.10 shows the edge of the foil being nicked with a pair of dykes, so it can be removed cleanly. The same action can be done by slicing the foil gently with a razor blade. It's helpful to grasp the blue mylar foil in one hand, the outer jacket in the other, and pull down the outer jacket so that more foil is exposed. Then nick the foil about {1/8} to {1/4} inch past the cut-off point of the outer jacket. When you release the outer jacket it will snap back, overlapping the foil cut and hiding any rough edges.

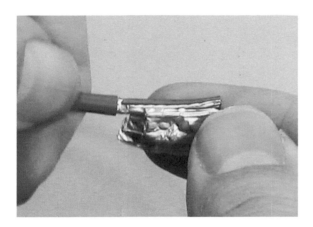

Figure 3.6.11 Removing the foil.

Pull on the foil where you nicked it and it will come away cleanly, exposing the wires (Figure 3.6.11). Clean up any rough edges by cutting them with dykes, and gently stretch the outer jacket back over the foil cut-off.

Separate the conductors so they can be worked on individually (Figure 3.6.12). Figure 3.6.13 shows high, low and shield (drain) conductors.

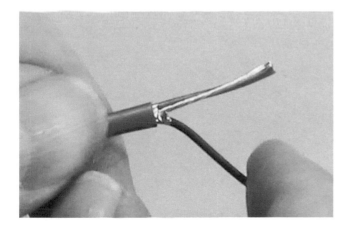

Figure 3.6.12 Separating conductors – 1.

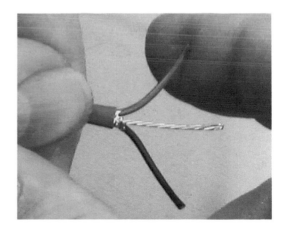

Figure 3.6.13 Separating conductors – 2.

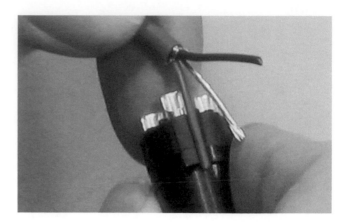

Figure 3.6.14   Length check by eye.

If you haven't wired many connectors, it's useful to hold the stripped wire against the connector to be soldered (Figure 3.6.14). This gives a hard-core reality check about how long the final length needs to be. When you're working from measurements, or have a clear understanding of the length needed, you can omit this step.

Figure 3.6.15 shows the correct length of a cut conductor – which is between {5/16} and {3/8} inch. Making the length longer than that reduces the physical strength of the connections.

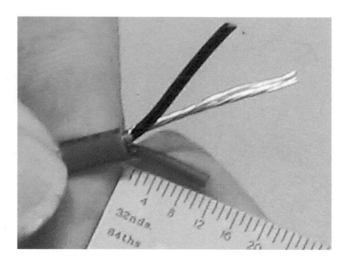

Figure 3.6.15   Cut conductors – 1.

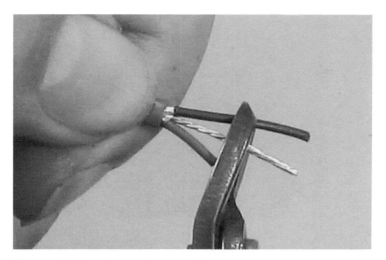

Figure 3.6.16   Cut conductors – 2.

Cut the other two conductors to the same length as the first one, between {5/16} and {3/8} inch (Figure 3.6.16). Try and get them all the same length.

All three conductors are nicely cut to be the same length (Figure 3.6.17). It gives you sort of a warm, fuzzy feeling, doesn't it?

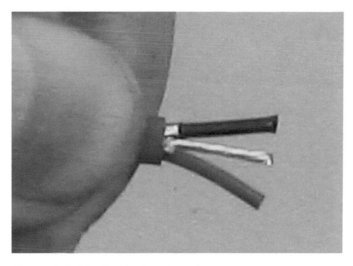

Figure 3.6.17   Finished cut length.

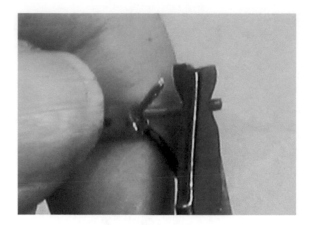

Figure 3.6.18   Stripping conductors.

Strip the ends of both the high and low conductors with a pair of wire strippers (Figure 3.6.18). I gauge the distance by the width of the jaws of the hand stripper, which is about {3/32} inch. If necessary, twist the strands of each conductor so they will stay in place to be soldered.

Both the high and low conductors have been correctly stripped to {3/32} inch, but the drain (shield) conductor is looking a bit frazzled (Figure 3.6.19). We fix that in our next step.

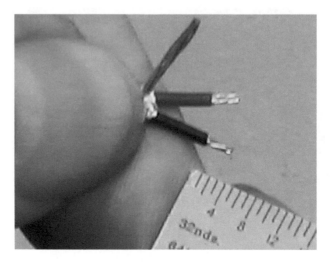

Figure 3.6.19   Correct strip length.

Twist the strands of the ground (drain) conductor together so they will stay in place to be soldered (Figure 3.6.20).

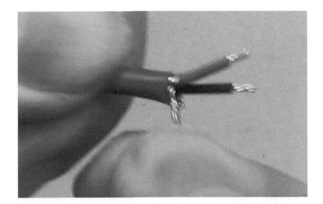

Figure 3.6.20   Twisting strands of drain conductor.

Mount the wire in a vise (Figure 3.6.21), and tin all three conductors. Mount it so the molten solder will flow (by virtue of gravity) toward the tip of the conductor rather than back onto the insulation. Make sure that your solder actually flows into the strands of the conductors and doesn't just coat the surface. Clean off any rosin blobs (once they've cooled) with your fingernail or a small screwdriver.

Figure 3.6.21   Mounting wire.

The actual tinning is shown in Figure 3.6.22. I only show the tinning of one conductor, but you'll remember to tin all *three*, won't you? Notice that my left hand is braced against the vise to steady it – similar to the brace you take for a pool (billiards) shot.

After tinning the wire, tin the solder cups of the connector (Figure 3.6.23). Fill them until they are almost brimming with solder, but not so full as to overflow when the wire is inserted. Remember that gently rounded spoon of sugar from Section 1?

Figure 3.6.22   Tinning wire.

Figure 3.6.23   Tinning solder cups.

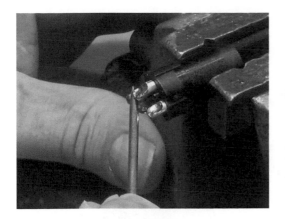

Figure 3.6.24 Scrape off excess rosin.

Figure 3.6.25 Filled solder cups.

Scrape off excess rosin from the solder cups with any convenient tool. In Figure 3.6.24 I'm using a small flat-blade screwdriver, but almost anything will do – even your fingernail.

Figure 3.6.25 is a lovely picture of the filled solder cups. All three show the correct amount of solder being added, and the chrome-bright appearance of good soldering.

Putting parts on the wire is a *very* important step (Figure 3.6.26)! If you forget it, you'll have to *un*solder the wires from the connector, put the plastic boot (left) and strain relief (right) onto the wire, and then *re*-solder the wire to the connector. Boring. So get it right the first time.

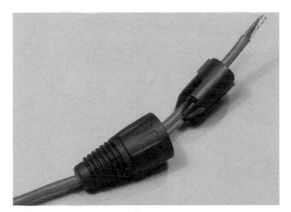

Figure 3.6.26 Putting parts on wire.

Figure 3.6.27   Attaching high conductor – 1.

Hurray! We finally get to attach the conductors. Note that I've flipped the iron to my left hand, so as to get a better shot at Pin 2 (Figure 3.6.27). Heat the solder cup on its top; heating will be too slow if you try from the bottom. When the solder is molten, quickly (but gently) slide the wire into the cup. Then gently slide the iron horizontally off – in this case, to the left – while holding the wire in place until the solder cools.

Also note that this shot is of a *female* XLR, with Pin 2 on the left. The male plug will have Pin 2 on the right.

Figure 3.6.28 is another view of the same action – Pin 2 on a *female* XLR. This one gives a better idea of the iron placement.

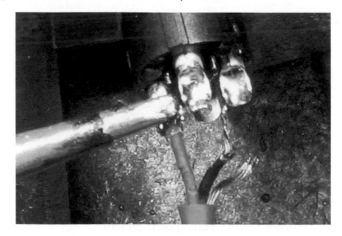

Figure 3.6.28   Attaching high conductor – 2.

To reinforce the point, Figure 3.6.29 shows Pin 2 being attached on a *male* XLR. Note that Pin 2 is on the *right* when the center pin is closest to you.

Figure 3.6.29   Attaching high on male.

Figure 3.6.30 shows the finished pin – a darn good Pin 2 solder job! Note that the insulation is flush to the cup, with *no* exposed strands. The cup is filled nicely, the solder is bright and shiny, and the length of *unshielded* conductor is very short, to reduce hum pick-up.

Figure 3.6.30   Finished Pin 2.

We are back to the female XLR in Figure 3.6.31, although for Pin 3 it doesn't matter. Pin 3, being in the middle, stays the same for both male and female XLR plugs. Isn't that considerate of it?

The procedure is the same as for the high (hot) conductor. Heat the cup, slide the conductor in, slide the iron off while holding the conductor in place, and wait until cool. Got it?

Figure 3.6.31   Attaching low conductor.

Cut back the drain conductor (shield) so it is a bit shorter than the high and low conductors, maybe {1/16} inch or so (Figure 3.6.32).

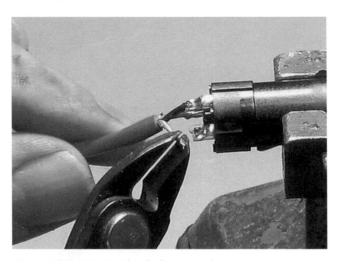

Figure 3.6.32   Cut back drain conductor.

As shown in Figure 3.6.33, this drain conductor has been delicately cut to {7/32} inch. Now we move on to the next step – soldering it (Figure 3.6.34).

Figure 3.6.33   Drain cut to length.

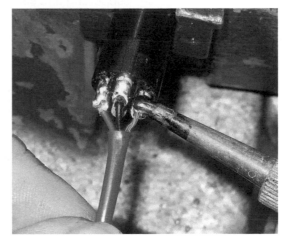

Figure 3.6.34   Soldering drain conductor.

It's a *female* XLR, so Pin 1 is on the *right*. Do the same ritual you did for the high and low conductors. Heat the cup, slide the conductor in, slide the iron off while holding the conductor in place, and wait until cool. It's easy!

We need a reality check here. If you're doing a *male* XLR, the Pin 1 cup will be on the *left* (Figure 3.6.35). But if you've attached Pins 2 and 3 correctly, there's no way you can mess up now, right?

Figure 3.6.35   Pin 1 on male.

That's right, all the soldering is done! See how short the exposed conductors are in Figure 3.6.36? They're so stubby and stiff that the wire is self-supporting and self-insulating, without any heat-shrink, electrical tape or other insulators that make doing the work slower. But there's a bit of exposed shield foil that we have to eliminate. It's the blue, shiny stuff that I've drawn the arrow toward.

Figure 3.6.36   Completed soldering.

Pull back on the outer jacket to expose more of the foil shield (Figure 3.6.37). How much foil you expose is not too critical, anywhere between {1/4} and {1/2} inch.

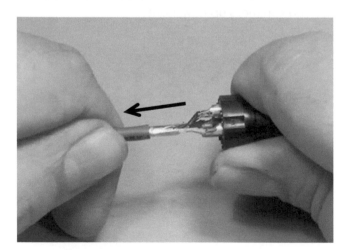

Figure 3.6.37   Pull back outer jacket.

Cut away {1/8} to {1/4} inch of the foil shield (Figure 3.6.38). That way, when the outer jacket is pulled back into position, there will be no way for the foil to short any of the XLR's pins.

Figure 3.6.38   Cutting back foil shield.

Our squeaky clean *female* XLR, ready to assemble, is shown in Figure 3.6.39. You've cleared up the foil, pulled the outer jacket back in place, and picked off any stray bits of rosin, insulation or other funky contaminants. Now to apply a contact enhancer and put the connector(s) together. Almost done!

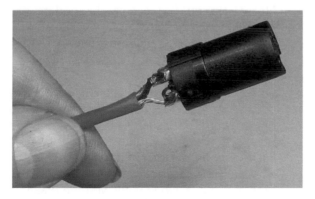

Figure 3.6.39   Female plug ready to assemble.

The male XLR being sprayed with enhancer is shown first (Figure 3.6.40), because that's the easy one to do. I'm using Caig ProGold G5 and *you* should too! It improves electrical contact on all metals used in electrical connections, and is especially good for low-power connections like microphone outputs and mating stage boxes. Most mics have very low power output – down in the microvolt range. At such a low level, any slight reduction in conductivity can be much more harmful than it would be at line-level or speaker-level. At higher power levels, the signal can cut through layers of (semi-conductive) corrosion.

Figure 3.6.40   Spray male with enhancer.

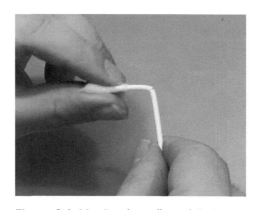

Figure 3.6.41   Break cardboard Q-tip.

Note: Caig Labs has recently changed the name from ProGold to DeoxIT Gold. It's the same stuff. They also have some new products that are interesting. More info about ProGold is on the Caig Labs website (www.caig.com).

One time I was feverishly searching for a way to clean dirty, old female XLR connectors. As an experiment, I broke a cardboard shaft Q-tip in half and tried the broken-off end of the shaft (Figure 3.6.41). It fitted like it was custom made to clean XLR females! This trick will *only* work with the cardboard shaft type or the wooden shaft type of Q-tip. It *won't* work with the plastic shaft ones; the plastic won't absorb the ProGold enhancer.

Saturate the broken end of the Q-tip with ProGold (Figure 3.6.42). What could be easier?

Swab the female connector's sockets with the wetted Q-tip shaft (Figure 3.6.43). Be sure to do all three and respray the Q-tip as needed. If the shaft end gets dirty, cut it back further with dykes.

Figure 3.6.42   Spray shaft of Q-tip.

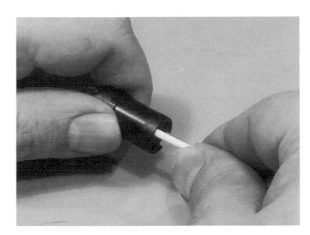

Figure 3.6.43   Swabbing female socket.

With new connectors, you can just spray the holes and wipe the surfaces. But when you're working with *old* connectors, this is the *only* way I've ever found to clean them. I've searched for years to find other ways. I'm still looking.

So remember this trick and some day you may be a rock 'n' roll hero, when you fix a crucial mic cord with a Q-tip and some rubbing alcohol.

All the components of the female XLR, ready to assemble with the metal outer shell, are shown in Figure 3.6.44. The male is very similar.

If the boot and strain relief are not correctly oriented, stop now and fix it, even if it means redoing the whole process. There's simply no other way. It's either right or it's done again, until it *is* right.

Figure 3.6.44   Ready to assemble – 1.

Figure 3.6.45   Ready to assemble – 2.

Figure 3.6.46   Final assembly.

Figure 3.6.45 is the same view as Figure 3.6.44, with the outer shell added. On to assembly (Figure 3.6.46)!

Push the plug and strain relief into the outer metal shell. Screw the plastic boot down to tighten the strain relief. These particular instructions apply to a Neutrik XLR connector, but other types are similar in use.

There's typically some type of strain relief, and a way to tighten/loosen it, a way to secure the plug to the shell, and a boot to help support the wire where it enters the connector. The completed connectors are shown in Figure 3.6.47.

That's it, gang. 'That's all she wrote' – or maybe I should say 'wired'. Now it's time to whip out that VOM you bought after I told you to do so in Section 1. If you really read the whole book, you'll recall a mini-course in how to use it (the VOM) that was part of Section 2. You'll even remember what VOM, or DVOM, stands for.

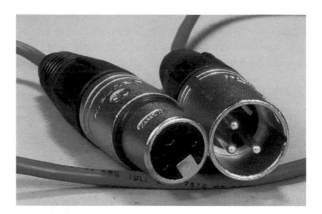

Figure 3.6.47   Completed XLR connectors.

Test your work, check it for shorts and high-resistance solder points. If it all checks out, enjoy your new connections!

TT male connectors

I can almost hear it – the question bubbling in all your brains, 'What the heck does the TT stand for?' There's actually an answer that makes sense; it stands for 'Tiny Telephone'. They are sometimes also called 'bantam' plugs – but I can't tell you where that comes from.

Remember those old telephone switchboards your grandmother worked on? They used a {1/4} inch diameter plug called a 'telephone' plug. Its construction was similar to a regular guitar plug, but it had a different shape for the tip of the plug, and also for the ring section.

The shape was different enough that guitar plugs and telephone plugs were not interchangeable. T-plugs don't make good tip contact in G-plug jacks, and putting a G-plug in a T-plug jack will stretch the contacts. But neither of those facts kept people from trying to put A into B, and vice versa. So there was a consistent stream of damaged connectors, often in difficult-to-repair places.

Old-school 'telephone' plugs had another, even bigger problem – they were too big! The maximum number you could fit in a 19-inch rack space was 24 or 26 connectors, depending on how tightly you spaced them together. As patch bays grew in size, especially with ever-larger track formats, the old plug style made patch-bay installations a floor-to-ceiling affair – not very workable.

Along came the design for the TT plug, and patch bays to match it. Since the TT plug is roughly half the size of a T-plug, bays could double in density for the same space. The old standard of two rows with 24 jacks each was now doubled, becoming two rows with 48 jacks each, in a 1.75 inch height (one RU – or rack unit) space. Patch bays became small enough to be mounted within recording consoles, where they were more convenient and ergonomic. And as an added benefit, there was no way that people could plug anything but a TT plug into a TT patch bay. Nothing else in the whole world fits.

But the smaller size of TT plugs and jacks makes them much harder to wire, and demands greater skill on the part of the wireperson. The TT is the most complicated of the soldered plugs I describe, so I've deliberately left it until last. If you're new to wiring, allow yourself extra time (and materials) to learn the skills you'll need for it.

I show the wiring of a TT male, as this is typically what you'd be wiring in the field. Patch bays (the females) are not within the scope of this book. If we sell lots of copies of the AWG, maybe we can do an 'advanced connectors' version. So tell all your friends to buy this book if you ever want to know about patch bays themselves.

The typical real-world situation is that you'd like to plug something into a TT patch bay – maybe a synth output, a guitar effects box or a mic preamplifier. Or you'd like to get an output from a TT bay for an audio feed to your video camera, to connect your mini-disk recorder input, or for a myriad of other reasons.

Either way, you're going from a (frequently) balanced audio realm – the patch bay – into other equipment that may be either balanced or unbalanced. I show the wiring of the TT for a balanced/stereo connection, i.e. tip, ring and shield. Your particular gizmo may be happy with a balanced TT to stereo guitar or whatever connector. Or your gizmo may require shorting ring to ground, or using only tip and shield. It varies from unit to unit, there's no overall rule.

If you wind up not using ring, or jumpering ring/low and shield, I suggest you do it at the gizmo end of your wire. Any other kind of plug is easier to wire – and *rewire* – than a TT.

Having thoroughly intimidated you with the difficulty of the TT connector, let me now show you how to master it – in only 68 easy steps!

Being such a tiny plug, there's a lot of info in Figure 3.7.1 – let's examine it. On the right-hand side we see the ball-shaped tip and then (moving leftward) the cylindrically shaped ring section, then the long barrel (shield) of the plug.

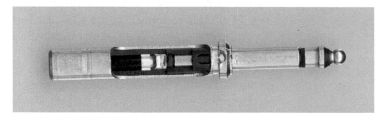

Figure 3.7.1   Plug with no sleeve.

In the middle of the plug, we have the black plastic (insulation) section, in which are two solder cups. The right-hand cup is for the tip, the left-hand cup is for the ring. There is *no* solder cup for the shield – we will have to create one later.

Those of you with eagle eyes will have noticed that the right-hand solder cup is convex, rather than concave. These kinds of things are known in the trade as 'design flaws'. They occur often enough for there to be a name for them. We can only wonder what the designer was smoking that day, and deal with the result. The lack of a solder cup for the shield is another design flaw.

Further, each manufacturer of TT plugs makes their flavor of the plug with a slightly different configuration, requiring slightly different techniques. I've chosen a Neutrik variation here, which is as typical as any other.

Figure 3.7.2 is just for reference – we won't be using the outer TT insulating sleeve, because we're using a large-diameter wire and heat-shrink instead of the sleeve. However, if you're using a smaller diameter wire, you might want to use the TT's sleeve for the outside covering. Small-diameter wire can

Figure 3.7.2 Plug with sleeve.

be 'built up' with several small layers of heat-shrink to match the internal diameter of the TT at the rear of the plug.

We'll come back to the TT plug very soon. First, let's play with some wire.

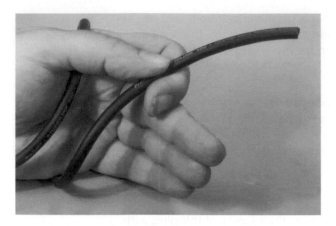

Figure 3.7.3   Wrap wire around hand.

Figure 3.7.4   Measure against plug – 1.

Wrap the wire firmly around your hand (Figure 3.7.3). This prevents the individual conductors from being pulled out when they are stripped, and is only needed on short lengths of wire. If the wire is already harnessed or strain-reliefed, you don't have to wrap it around your hand.

Since different makes of plugs will have different lengths, you should measure the amount of outer jacket to be cut away against the plug itself (Figure 3.7.4). With this particular plug, we get a length to cut back of {1.375} inch. We're stripping back a little more then we need to, so we can have some slack to play with. Note that by placing my thumbnail at the point being measured, I can guide my stripping tool safely to the exact spot I want. This makes it easier to strip back an accurate distance each time.

Still holding my thumb in position, I verify that {1.375} inch strip length (Figure 3.7.5). You could also call it a {1 and 3/8} inch strip, if that's easier.

Figure 3.7.5   Measure against plug – 2.

The wire I'm using for this example is two-conductor shielded Monster, one of the highest quality types available. Like many larger diameter wire types, it has a thick, rubbery outer jacket, which cuts more cleanly with a razor blade than with a pair of wire strippers (Figure 3.7.6). However, should you prefer, stripping the outer jacket off with wire strippers is also fine.

If using a blade, gently rock it at a right angle to the wire and then, using a light slicing motion, spin the blade around the wire, keeping it at a right angle at all times. This technique will create a super-clean cut of the outer jacket and, if done gently, will not harm the strands of the shield conductor that lie just inside the outer jacket.

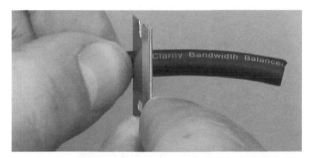

Figure 3.7.6   Cut outer jacket with razor blade.

Figure 3.7.7   Break open razor cut.

Gently bend the cut in all directions to break away any slivers of the outer jacket that still connect both sides (Figure 3.7.7). If needed, cut the slivers lightly with the razor blade. The break should be complete on all sides.

This particular wire is very supple, and the outer insulation jacket can just be pulled off (Figure 3.7.8). One of my crude arrows helps illustrate the direction to pull. If the outer jacket is sticking to the shield, then you have to cut it lengthwise (see the details in Figure 3.1.8). I'm breaking my own rule about flipping – but this wire behaved itself and didn't need the additional cut.

Figure 3.7.8   Pull end off.

This particular Monster shield is not tinned, and is loosely woven, so it's flexible and easy to unbraid, unlike some other shields. I was good – I didn't press too hard with the razor blade – so none of the strands are broken.

Using any convenient small pointy object, carefully unbraid the shield strands (Figure 3.7.9). Do this one or two overlaps at a time; don't try to do a whole bunch at once – you'll break strands. As you can see in the picture, the awl blade on a Swiss Army knife is ideal for this operation. Note that I'm using the blunt edge of the awl to pull through the strands – there is less chance of breaking them that way.

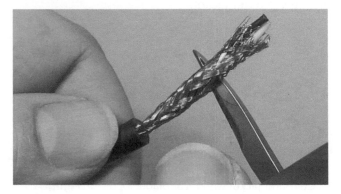

Figure 3.7.9   Unbraiding shield – 1.

Keep going, keep going – it won't take long. Work all the braiding out of the shield strands, right down to where they break out from the outer jacket (Figure 3.7.10). This particular wire is very civilized, with no fuzz, string or plastic wrapper – just the three conductors. You might not be so lucky – be prepared for a little clean-up. Trim all leftover insulation as tightly as possible, without harm to the conductors.

Figure 3.7.10   Unbraiding shield – 2.

Hold the wire in your hand (or a vise), and smooth out the strands of shield so they lie flat (Figure 3.7.11). The goal is to make a thin, broad surface to solder onto the strain relief. Think of a strip of paper lying limply – not a twisted rope, as one might be tempted to make.

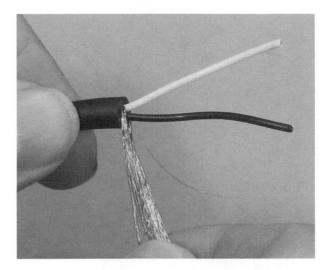

Figure 3.7.11 Smoothing out shield.

Now that you've got those shield strands lying obediently flat, you need a trick to keep them that way. At the end of the shield strands, twist them around each other in a circular motion (Figure 3.7.12). This will leave the bulk of the strands lying flat, and flexible, but keep the whole bundle of strands together for soldering.

Figure 3.7.12 Twist shield at end.

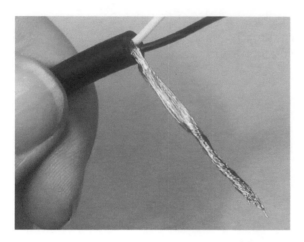

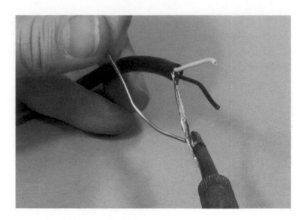

Figure 3.7.14 Solder twist in shield.

Figure 3.7.13 Twist shield detail.

Figure 3.7.15 Shield solder detail.

Figure 3.7.13 is a fly's eye view of the twist at the end of the shield. Now we need some way to keep it twisted – which brings us to Figure 3.7.14.

Solder the shield conductors at the twisted point. Use only enough solder to keep the strands together at the twist. Don't let the solder run into the flat portion you created – that would make the shield stiff, when we need it to be flexible.

See in Figure 3.7.15 how the soldered area is localized to the twisted strands? This leaves the rest of the shield flexible for later work.

In Figure 3.7.16 I'm cutting the end of some layout tape (also called 'artist's tape') so it will lie flat when I wrap it around the exposed conductors. The next operations are somewhat violent, and one mis-cut could damage the delicate conductors.

Measure a little past the ends of the conductors, and cut the layout tape (Figure 3.7.17). This is not a critical length, just cover the conductors. Then roll the layout tape around the conductors to create a temporary protective covering. Think of a lobster-bib for wire.

The well-wrapped (and protected) conductors are shown in Figure 3.7.18. They are able to withstand minor abuse in assembly. The layout tape comes off easily when no longer needed.

Figure 3.7.16   Cover conductors – 1.

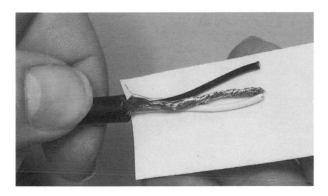

Figure 3.7.17   Cover conductors – 2.

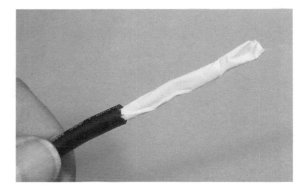

Figure 3.7.18   Cover conductors – 3.

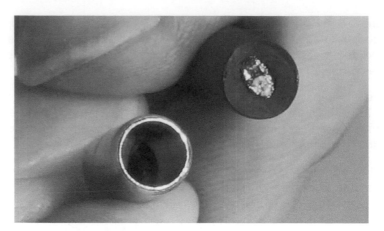

Figure 3.7.19 Wire vs. plug diameter.

In Figure 3.7.19 we see one of the reasons for the layout tape: the diameter of the outer jacket on the Monster wire is larger than the diameter of the wire entry point at the rear of the TT plug. Even worse, the spacing on TT patch bays is so tight that more than a tiny buildout on the plug will interfere with plugs next to it. Well, if you can't raise the bridge, lower the river. We *can* make the outer jacket fit if we carefully carve it with a sharp razor blade, to reduce its diameter.

The next question is how far back to carve the outer jacket? By comparing the plug with the jacket, I can see that carving back to the zero to the left of the dot, just left of the word 'Ultra', will give me enough carved length for a good fit (Figure 3.7.20).

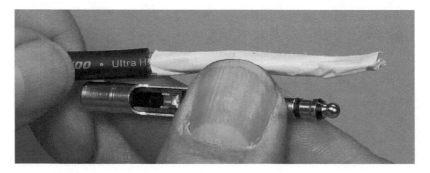

Figure 3.7.20 Length to carve back.

Figure 3.7.21 Carve jacket – 1.

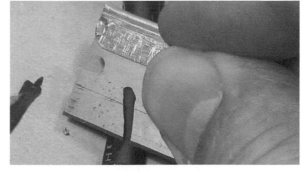

Figure 3.7.22 Carve jacket – 2.

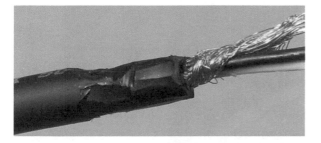

Figure 3.7.23 Carve jacket – 3.

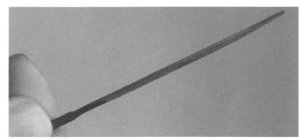

Figure 3.7.24 Triangular needle file.

In Figure 3.7.21 I've got the wire lying flat on the table, the razor blade is cutting *away* from my fingers, and the blade is lying flat against the wire, to help guide the cut. I'll cut this section and then rotate the wire to do another. I'll keep going until I've gone all the way around the outer jacket.

In Figure 3.7.22 I've shifted the position of my hands, but the rules are the same – cut away from your own fingers and cut many little slices, rather than a few big ones.

The finished carving is shown in Figure 3.7.23. No one will hire me for fine sculpture, but it's great for a TT plug. Notice that I've also removed the layout tape before inserting the conductors into the plug, as we do soon.

It's time to put the wire aside – it's as ready as it can be, until we modify the TT plug itself. Our first job is to create a solder cup for the shield – or maybe a 'solder channel' would be a better description. To make such a miniature solder channel, we need a small, precise tool. Hence my introduction of the triangular needle file, which I will forever after call a 'TNF' for brevity (Figure 3.7.24). The TNF typically comes in a set of needle files, available at your local hardware store. No substitutes are allowed – nothing else will work as well.

In Figure 3.7.25 I've put the plug in my vise, at an angle that allows for easy filing. Notice that the file is almost at a right angle to the top of the plug (the area with the solder cups). This permits me to file a broad groove/channel with the TNF.

The completed soldering channel is shown in Figure 3.7.26. Notice that it broadens out? You might say I really 'got in the groove' for this work – but we still have to tin all the solder points, which comes next.

Now that I've created that nice, broad solder channel I must tin it, as I want the solder of the finished joint to meld quickly and cleanly (Figure 3.7.27).

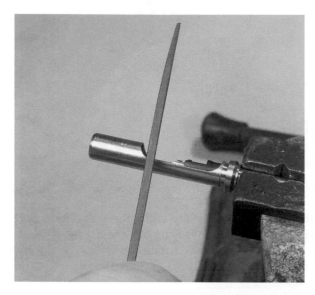

Figure 3.7.25   File solder channel.

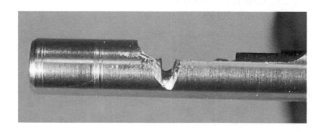

Figure 3.7.26   Filing completed.

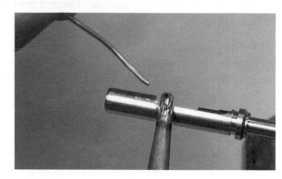

Figure 3.7.27   Tin shield cup.

Unlike almost every other tinning operation, we don't mind a little overflow here (Figure 3.7.28). All the excess will be filed away shortly.

An extreme close-up of our notorious (tinned) shield channel is shown in Figure 3.7.29. Notice that I've filed past the metal of the plug, into the black insulation section. This allows the shield strands to feed directly out at a right angle.

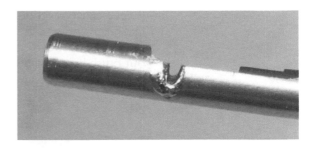

Figure 3.7.28   Tinned shield cup – 1.

Space is so tight inside the connector that I want to pre-tin the high and low conductor solder cups. This way, when I bring the solder-laden conductor into position against the solder-encrusted cup, they will bond immediately when heated. That's the premise, at least. In Figure 3.7.30 I'm tinning the high cup.

Tin both cups the same way – with a thick film of solder, enough to provide 'wetting' action against the solder on the conductors, but not so much as to drip down and short. In Figure 3.7.31 I've tinned the cup, and am starting to gently pull away the iron tip. You can see the solder sticking to the tip; as I pull away further, the solder will fall back into a nice round blob (I hope).

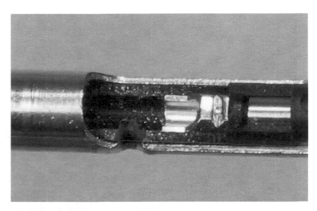

Figure 3.7.29   Tinned shield cup – 2.

Figure 3.7.30   Tinning high cup.

Figure 3.7.31   Tinning low cup.

Finally, all three solder cups are tinned and ready (Figure 3.7.32). The black insulation shows a little melting, but there are no voids or solder bridges. Brush off any loose particles. Leave any rosin you find on the solder cups, but brush it away from other areas.

Figure 3.7.32   Tinning completed.

Insert the wire at the rear of the plug (Figure 3.7.33). Pull the conductors toward the tip of the plug to avoid them wrapping around each other. Push the outer jacket as far into the rear of the plug as it will go by hand.

Using a pair of pliers, twist (rotate) the plug down onto the jacket (Figure 3.7.34). Do this about a quarter turn at a time, checking the conductors and untangling them, if necessary. There are two objectives – to firmly seat the jacket in the plug, and also to line up the shield conductor strands with the solder channel we made for them.

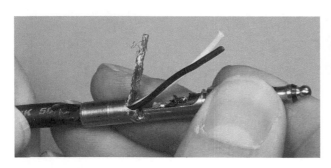

Figure 3.7.33   Insert wire.

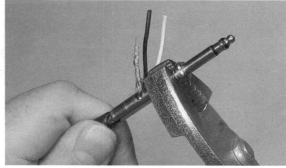

Figure 3.7.34   Twist plug onto jacket.

The shield strands must go directly to the shield solder channel, and not overlap the high and low conductors. This is clearly shown in Figure 3.7.35.

The figure shows the wire correctly positioned. See how the shield strands all go directly to the shield solder cup?

Figure 3.7.36 is the same as Figure 3.7.35, but from a slightly different angle. All the shield strands flow neatly into the shield solder channel. Now to cut off the excess shield.

Figure 3.7.35 Wire in position – 1.

Figure 3.7.36 Wire in position – 2.

Snip! I just cut off the extra shield strands – I did it so fast you didn't even see it. Alright, I'm fibbing – I missed a shot. But the result can be seen in Figure 3.7.37; the strands are cut close to the barrel of the connector, but with some excess still in place.

Since the solder point of the shield to the barrel is mainly what's going to hold this wire in place, we want to be generous with solder here. The shield strands will absorb a lot, making them stronger. Notice that I'm soldering almost vertically in Figure 3.7.38. That way, any overflow will spill down the outside of the connector barrel rather than into the plug itself. In fact, a bit of blobby overflow is actually good here, since it will be filed to conform.

Figure 3.7.37 Excess shield cut.

Figure 3.7.38 Solder shield strands.

As can be seen in Figure 3.7.39, the shield strands are well tinned, and there's a gorilla-strong connection between the shield and the barrel. But what about our old friend there, 'blobby overflow'? Not to worry – they make *big*files too!

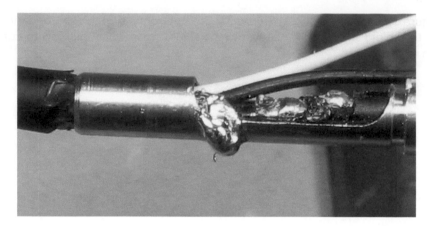

Figure 3.7.39   Completed shield solder.

There's nothing like a big, old, metal file to take care of that pesky overflow (Figure 3.7.40). File gently, with a curving, circular motion. The goal is for the curvature at the solder point to match the curvature of the plug barrel itself.

Figure 3.7.40   File shield solder.

Figure 3.7.41   Completed filing – 1.

Figure 3.7.42   Completed filing – 2.

Neat! The curvatures match exactly (Figure 3.7.41). Now we can use either heat-shrink or a sleeve for the outside insulation of the TT plug.

Figure 3.7.42 shows the same detail from a different angle. See how the shield strands are filled with solder, but none has flowed over onto the other parts of the plug? I'll brush off the detritus, and start finalizing the high and low conductors.

I wish I could give you a fixed length to cut – but I can't. It will vary with each connector you wire. In Figure 3.7.43 I'm using black for the low conductor, so I've cut it to match the distance to the low solder cup, with a smidgen more for a little curve.

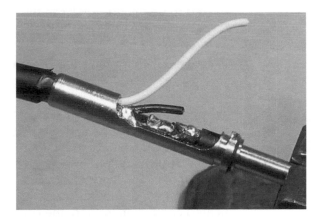

Figure 3.7.43   Cutting low conductor.

Figure 3.7.44   Stripping low conductor – 1.

Figure 3.7.45   Stripping low conductor – 2.

In this case, both the white and black conductors are electrically identical. So you can choose either one for the low or high, but you *must* be consistent! If black is high at one end of your cable, make sure it's high at the other end too. Electrical convention in AC wiring is that black is low, but you're not bound by law to follow that – only to be consistent in your work.

Since the TT is so small, the low conductor must be soldered in place before we can do any work on the high conductor.

Figure 3.7.46   Tinning low conductor.

Strip the low conductor (and the high one too) back {1/16} inch (Figure 3.7.44). Why such a small amount? Because the insulation on cable tends to shrink (or 'wick') back when the strands are heated to be tinned. You'll see this in the next steps.

In Figure 3.7.45 it seems too small a strip, but wicking (melt-back) will increase it greatly.

We need to use a technique called 'beading' here, where the tinned ends of the conductors are loaded with extra solder (Figure 3.7.46). This forms (ideally) a grape- or bead-like shape – hence the name. This technique is used where it is not possible to feed the solder onto the heated conductor. Instead, the conductor itself must carry the solder as a 'payload'.

The first step in beading is to tin the end of the conductor. Sometimes it is possible to add the full payload of solder all at once. Often, however, it helps to let the tinning cool down, and then add the beading (build-up) as a second step. The following pictures show this process.

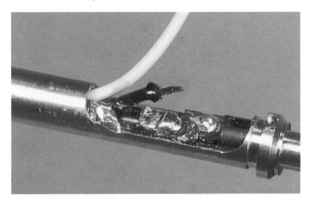

Figure 3.7.47 Tinned low conductor.

Notice in Figure 3.7.47 how much the insulation has wicked back? We still have to add the bead of solder in a second operation – but let it cool for a bit first.

We are back again, with a fuller payload of solder, in Figure 3.7.48. Melt just enough solder to form a bead at the end of the wire. Pull the iron away and let it cool undisturbed, or blow gently on it.

See the nice, shiny, round ball of solder in Figure 3.7.49? That's a good bead, ready to melt quickly. Now we have to position it close to the low solder cup.

Figure 3.7.48 Beading low conductor.

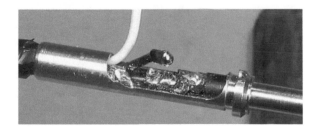

Figure 3.7.49 Low bead completed.

Remember that curve I told you I wanted in the low conductor? Here it is in Figure 3.7.50, and the bead is resting in position on the low solder cup. Now I'll press gently down with my finger on the insulation of the low conductor, and at the same time put my iron tip directly on the cup and solder bead. This is the 'before' shot with the bead still discrete.

Figure 3.7.50   Low bead positioned.

Figure 3.7.51 is the 'after' shot, with the bead soldered in place. There is a strong physical connection, and a slight arc in the conductor for flexibility. No, I didn't show you the naked act of soldering itself – but I'll get another chance real soon.

Figure 3.7.52 is another 'after' shot, from a different angle. See how tight the connection is between the low conductor and the low solder cup?

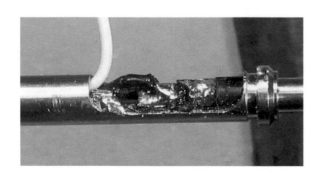

Figure 3.7.51   Low bead soldered – 1.

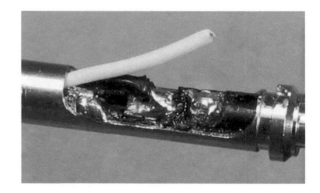

Figure 3.7.52   Low bead soldered – 2.

Darn, I missed another shot here – I should show the high conductor being cut to length. And again, you have to measure the length yourself – each plug will be unique. Will you accept as logical that we have to cut to length before stripping? That I did so for us? Good! Then let's strip this conductor back {1/16} inch, just like the low one (Figure 3.7.53).

Once again we have a strip of {1/16} inch, which will increase when tinned (Figure 3.7.54).

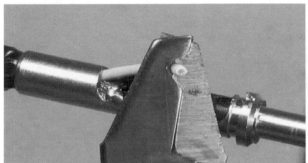

Figure 3.7.53   Stripping high conductor.

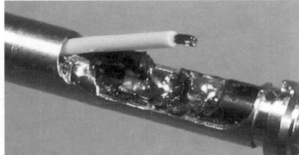

Figure 3.7.54   Stripped high conductor.

As for the low conductor, this is a two-step process. Tin first, allow to cool, then bead (Figures 3.7.55 and 3.7.56).

Wow! This one happened to bead perfectly in one operation. Sometimes you just get lucky. If you're not so lucky, do the tin and bead in two steps.

Figure 3.7.55   Tinning high conductor.

Figure 3.7.56   Beaded high conductor.

Remember that curve I told you I wanted in the low conductor? It's back!
I want it for the high conductor too. The bead is resting in position on the
high solder cup. I'll press gently down with my finger on the insulation of
the high conductor, and at the same time put my iron tip directly on the
cup and solder bead. Figure 3.7.57 is the 'before' shot with the bead still
discrete.

Figure 3.7.57   Position high conductor.

Ah! Ken and I finally got it together to show you that molten moment of
melding, when the iron tip hits the solder bead and the tinned cup at the
same time (Figure 3.7.58). Watch your fingers but if you do this right, 2–3
seconds of iron contact should be all that's needed.

Figure 3.7.58   Solder high conductor.

The completely wired TT plug is shown in Figure 3.7.59. Now all it needs is some outer insulation. The factory sleeve won't fit over the wire – time for some heat-shrink!

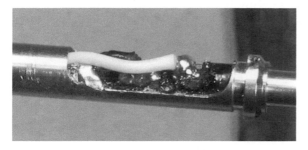

Figure 3.7.59 Soldered high conductor – 1.

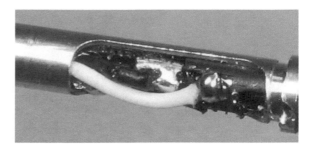

Figure 3.7.60 Soldered high conductor – 2.

Figure 3.7.60 shows the same detail from a different angle. See how the high conductor curves around the low conductor to keep a low profile? It's important to keep the conductors as flat as possible for this type of plug – they're packed so tightly when in use.

Now I need some heat-shrink big enough to just fit over the plug. Ah, got some, and it's clear, so you can see the details inside it – good! We want a long overlap past the point where the wire enters the plug (Figure 3.7.61). This overlap will help keep the wire in place, and adds durability. The exact length is not critical.

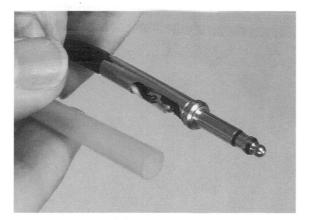

Figure 3.7.61 Sizing heat-shrink.

In Figure 3.7.62 the first layer of heat-shrink is in place but has not shrunk down yet. The shrinking process is taking place in Figure 3.7.63. Yes, I know it's blurry – but everything's moving here! I'm rotating the wire to distribute the heat, and also running back and forth along the length of the plug with the heat gun, to further distribute the heat. I have to do this to get a smooth, even shrink. So please be tolerant of minor imperfections.

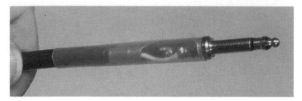

Figure 3.7.62 First layer of shrink.

Figure 3.7.63 Shrinking first layer.

Figure 3.7.64 shows the first layer of heat-shrink, after heating. Looks good, but I think I want a second layer for reinforcement. Heck, that's easy now.

Figure 3.7.65 shows the second layer of heat-shrink in place, but not shrunk yet.

Here we go again. Rotate the wire, move back and forth with the heat gun. Nice smooth shrink down. Figure 3.7.66 is blurry too – it's all still moving.

You probably thought we'd never get there, but here it is in Figure 3.7.67, a finished male TT connector, ready to work for many years. But there's one last touch.

Figure 3.7.64 First layer shrunk.

Figure 3.7.65 Second heat-shrink layer.

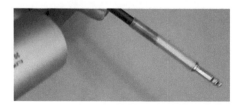

Figure 3.7.66 Heat second layer.

Figure 3.7.67 Finished TT connector.

Figure 3.7.68 Spraying with ProGold.

Figure 3.7.68 shows the outside of the connector being sprayed with enhancer. Wipe off any overspray. I'm using Caig ProGold G5, and recommend that you do too! It prevents aerobic corrosion – which degrades conductivity. It improves electrical contact on all metals used in electrical connections, and is especially good for low-power connections like guitar outputs and mating stage boxes. Most guitars have very low power output – down in the microvolt range. At such a low level, any slight reduction in conductivity can be much more harmful than it would be at line-level or speaker-level. At higher power levels, the signal can cut through layers of (semi-conductive) corrosion.

Note: Caig Labs has recently changed the name from ProGold to DeoxIT Gold. It's the same stuff. They also have some new products that are interesting. More info about ProGold is on the Caig Labs website (www.caig.com).

That's it, folks. 'That's all she wrote' – or maybe I should say 'wired'. Now it's time to whip out that VOM you bought after I suggested you do so in Section 1. If you really read the whole book, you'll recall a mini-course in how to use it (the VOM) that was part of Section 2. You'll even remember what VOM, or DVOM, stands for.

Test your work, check it for shorts and high-resistance solder points. If it all checks out, enjoy your new connections!

3.8

BNC male connectors

The BNC connector is commonly used for video and RF work. It's an unbalanced connector, with only hot (tip) and ground (shield). Unlike its poor cousin, the F connector (covered in Section 3.9), the BNC uses a spiffy gold-plated pin for the tip contact. And rather than screwing down on the female with a lock-ring you have to spin and spin, the BNC has a twist-locking lock-ring that only takes a quarter turn to lock firmly.

These features make the BNC a better connector than the F connector if you have a choice of which type to install. It's a little more costly, but has much better longevity. It's also a little more complicated to assemble than an F connector, but not a whole lot.

You can't say that BNCs are soldered; instead they are crimped (carefully crushed) and, like the F connector, you need a good quality crimper to do the work. Buy, borrow or rent a heavy-duty metal crimper – ideally one with interchangeable jaws, so you can use it for both BNC and F crimping. Cheap, flimsy, plastic crimpers will ruin a lot of connectors for you.

Let's take a look at a finished BNC (Figure 3.8.1), to see the final product, before we go through the assembly steps.

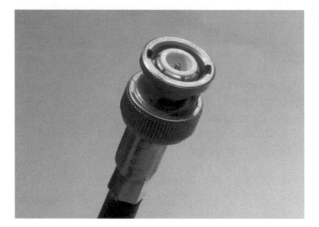

Figure 3.8.1   Finished connector.

The especially alert among you will realize that this finished example is slightly different from the BNC we'll be wiring. My point in showing it is that there are different types available, and the parts from one type may or may *not* fit another type. So keep the parts for each model of BNC together, or you'll be creating extra work for yourself.

Oh, and just to compound your confusion, BNCs come in two flavors: 50 and 75 ohm. Figure 3.8.1 is a 50 ohm, as it has a plastic insulator. If there was no insulator, it would be the 75 ohm version.

You can find a good discussion of which to use, where and why, at this URL: http://www.l-com.com/content/Tips.aspx – look in the 'Coaxial' section. Since the crimping of both types is identical, I'm only going to show you the assembly of one type of BNC.

Looking at our example, we can see the free-spinning lock-ring, the center pin and, on this particular plug, two different levels of crimp at the rear of the connector, where the wire enters. There's a smaller diameter crimp on the wire itself, and a larger crimp around the body (barrel) of the shell. This is unusual; most BNCs use only one crimp diameter. But it's important to understand that different models of BNC will require slightly different techniques. I can't show all the variations, so I'll use the most common type and ask you to adjust your procedures if you work on different style BNCs.

The front of the BNC is shown in Figure 3.8.2: on the left, we see the crimp ring; on the right, the body of this BNC connector and, in the center, a lovely, gold-plated center pin. Notice the center hole in the connector body. The pin is crimped to the center (hot) conductor of the wire, and then pushed through the center hole to seat it.

Figure 3.8.2   Front of BNC.

See the split cylinder inside the lock-ring? Its function is to slide around the (metal) outer layer of the BNC female plug, making a fast, high quality electrical connection for the wire's shield (ground).

The rear of the BNC is shown in Figure 3.8.3. What's different here? Not the crimp ring, not the center pin. Rather, I've turned around the connector body, so you can see the entry point for the wire. After the wire and center pin are in place, and the shield has been pushed up and over the cylinder that the wire enters, the crimp ring is put in place and crimped down. We'll work our way through all these steps sequentially; I just want to give you a quick overview.

Figure 3.8.3   Rear of BNC.

Figure 3.8.4   Two coaxial wires.

Two typical examples of coaxial wire are illustrated in Figure 3.8.4. Note the single inner conductor, with a thick insulator around it. Also note that the wires are of different diameter, so they need BNC connector shells with different diameters to fit them properly. We'll talk about that more in a minute, but for now I want you to observe that the wire on the left has only a braided shield conductor, but the wire on the right has an additional foil shield. This provides more complete shielding, at a bit more cost, and creates a stiffer wire.

We'll be working with the single-shielded version, as it's the one that fits the BNC plugs I have at hand. Working with the double-shielded wire only adds the foil, which is treated as part of the braided shield.

Do unto the foil as you do unto the braided shield, at each step, and you'll be fine. You can also sneak a quick look at the next section (3.9) for some pictures of foil-play.

BNC connectors and wire come in various sizes and standards. There's the older RG-59U, the thinner (newer) RG-6U, and a bewildering array of other types and sizes. The important thing to check is that the diameter of the inner insulator and the diameter of the wire entry tube of the BNC connector are matched. You shouldn't take some clerk's statement that 'They'll fit perfectly' as truthful – the clerk might not know. Peel back a bit of the shield and outer jacket, and physically check the fit of the inner insulator to the wire entry point of the BNC connector shell.

While it is possible to strip wire for BNCs by hand, with a razor blade, the cutting is complicated enough that a specific BNC stripper will save you a *lot* of time. I do, however, give exact cut lengths later, so you can do the prep work with only a razor blade and some careful workmanship.

So 'meet my little friend' – the BNC stripper (Figure 3.8.5). The orange ring at the bottom of it pushes two rollers up, and forces the wire against three (very sharp) blades. By setting the blade depths carefully, you can basically do the entire strip in one quick series of operations. Let's look at some details.

Figure 3.8.5 BNC stripper.

The free-spinning rollers push the wire progressively up, so the blades cut deeper each time you rotate the stripper around the wire (Figure 3.8.6). The orange cut depth ring has click stops, so you can advance it a little bit each time you spin the stripper around the wire. By cutting the wire deeper in small increments, you can get a very clean, accurate cutaway. Notice the graphic on the stripper at the top? It shows a stylized example of what a correct strip should like, with each layer of the wire removed an appropriate length.

Figure 3.8.6   Stripper rollers.

Figure 3.8.7 shows a close-up of the three stripper blades, set to increasing depths. They are set with adjustment screws – which brings us to Figure 3.8.8.

Figure 3.8.7   Stripper blades.

Figure 3.8.8   Blade adjusters.

Logically enough, the three adjustment screws are at three different depths, to set the blades. You have to dedicate a few inches of coax wire to setting the blades for the exact wire batch you're working with. Each batch of wire will have a slightly different blade setting for an optimal cut. Experiment until you get exactly the cut you need for the wire you're working with; one size does notfi t all.

CDR is my shorthand for the orange 'cut depth ring', which is way too long a name to say too often. In Figure 3.8.9 I'm rotating it up toward the blades, leaving just enough space to lay in my wire. I've drawn a couple of crude arrows to illustrate the direction of rotation.

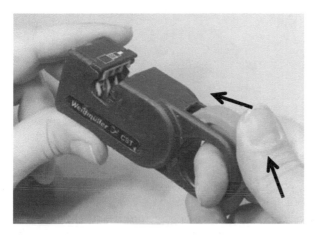

Figure 3.8.9   Adjust CDR – 1.

Figure 3.8.10   Adjust CDR – 2.

In Figure 3.8.10 I've pushed the CDR up so the wire is resting against both the rollers, and the blades, but no cut has been made yet. Next, we'll start advancing the CDR and spinning the stripper around the wire several times. Each type of wire will have a different resistance to the cut, and will need a different combination of CDR advance and stripper spin. You just have to 'feel it out' for the wire you're working with.

Figure 3.8.11 shows our first time around the wire with the stripper. The arrows give a rough idea of the rotation. You can actually *feel* the blades slicing through the braid and insulation. Use this as a guide to how many rotations are needed before the next CDR advance. When the resistance is gone, it's time to advance the CDR another click, and then spin around again.

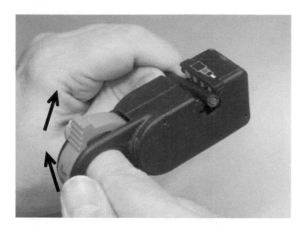

Figure 3.8.11   Spin stripper – 1.

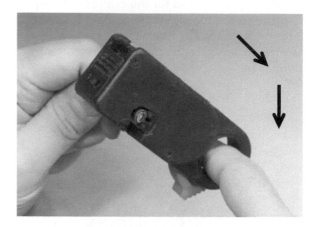

Figure 3.8.12   Spin stripper – 2.

Note the ridged part of the CDR in Figure 3.8.11 – the place you'd push against with your thumb to advance it. Before advancing, it was down at the base of the stripper. Here it's about halfway up, and we'll keep advancing it a bit with each rotation. Each time, of course, the blades cut deeper, a little at a time.

Figure 3.8.12 shows the back of the spin, with some more arrows to reinforce the concept. Once you get the hang of it, and the blades are set correctly, this series of actions can be done very quickly, with the entire strip taking less than a minute.

Notice the CDR position in Figure 3.8.13 – I've advanced it a couple of clicks, and now I get to take another spin, as it were.

Figure 3.8.13   Spin stripper – 3.

In Figure 3.8.14 I've advanced the CDR to its final position – the blades are at their deepest cut. One or two more spins and I'm done.

Figure 3.8.14   Spin stripper – 4.

Our multiply-cut wire is shown in Figure 3.8.15. Now it's time to remove all the cut-off braid and insulation. The next series of pictures illustrates this.

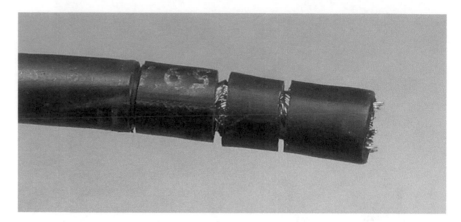

Figure 3.8.15   Finished cut.

In Figure 3.8.16 the first bit of cut-off is removed. But here, we want the inner conductor, right? Just stay tuned.

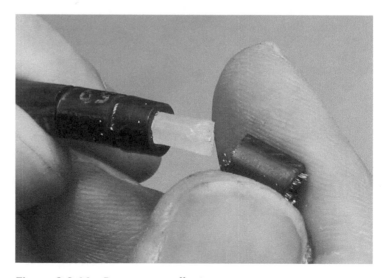

Figure 3.8.16   Remove cut-off – 1.

In Figure 3.8.17 the inner insulation has been deftly sliced by the blade with the deepest cut, and can be easily pulled off. In fact, as shown in Figure 3.8.18, both the inner insulation and the cut-off from the next layer came away cleanly.

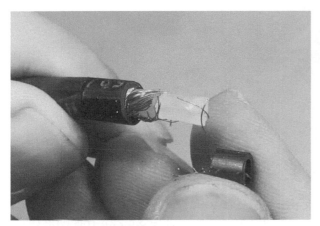

Figure 3.8.17   Remove cut-off – 2.

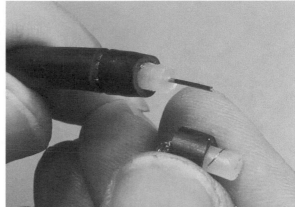

Figure 3.8.18   Remove cut-off – 3.

In Figure 3.8.19 the braid was not cut as cleanly; we have some strands that are longer than those that were cut by the stripper's blade. We have to fix this, and justify all the braid strands to be the same length (Figure 3.8.20). Cut back all the strands to be the same length as those cut by the stripper.

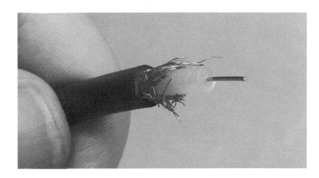

Figure 3.8.19   Remove cut-off – 4.

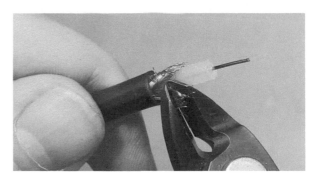

Figure 3.8.20   Justify strands.

The finished strip is shown in Figure 3.8.21. We've got {1/4} inch of neatly trimmed braid, another {1/4} inch of inner insulation, and {1/8} inch of bare center conductor. The masochistic among you might try to duplicate this with a razor blade. It can be done, but it's a lot slower.

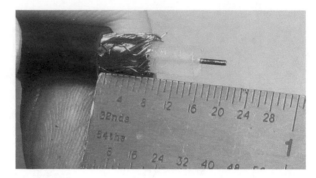

Figure 3.8.21   Finished strip.

It's now time to crimp the center pin. But wait, for that we need a (good) crimper, right? We certainly do, and here it is (Figure 3.8.22)! This type of crimper can be used for a variety of plugs, including BNC, F and Ethernet connectors. Figure 3.8.23 shows different jaw-sets; the ones mounted in this picture are for BNC connectors. The Ethernet set is on the lower left, and the F connector set on the lower right. This crimper is versatile, durable, and has a powerful set of jaws! It's got a crushingly powerful personality and delights in crunching various plugs into obedient conformity.

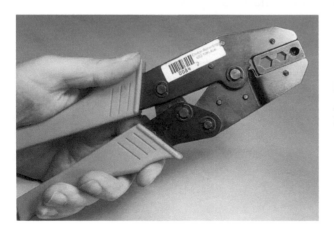

Figure 3.8.22   The crimper.

Figure 3.8.23   Multiple jaw-sets.

Speaking of those jaws, a crimper will only release after full compression, so *avoid getting any part of your anatomy caught in the jaws.* You will either have to let that part get squished, bloody, and mangled, or take the crimper apart to release yourself! The same goes for anything else you put in the crimper, so be careful! Powerful tools are always dangerous. Use them with respect and good judgment.

A close-up of the BNC jaws is shown in Figure 3.8.24 – note the different sizes for both the center pin and the crimp ring. Always start out with the larger size and work down, to avoid jamming the crimper.

Figure 3.8.24   Close-up of BNC jaws.

In Figure 3.8.25 I'm using the thumbscrews to mount the BNC jaws, so we can get down to some crimping. In Figure 3.8.26 the crimper is now set to play – let's start!

Figure 3.8.25   Mount BNC jaws.

Figure 3.8.26   BNC jaws ready.

I've carefully trimmed back the center conductor, so it runs the full length of the space in the rear in the center pin, but still allows the base of the center pin to rest on the inner insulation for mechanical strength (Figure 3.8.27). Note the little round hole in the crimpable area of the center pin? See the center conductor through the little hole? That's what the hole is there for – so you can look and see if the center conductor is in position, and long enough.

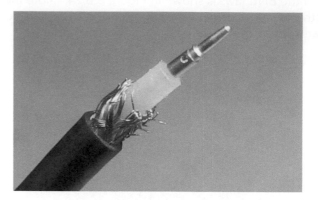

Figure 3.8.27   Mount center pin.

Carefully crimp the center pin (Figure 3.8.28). Start with the larger diameter crimper and work down, if needed. Make *sure* to crimp on the rear of the pin, and do *not* crimp on the front shoulder of the pin. If you do that, the shoulder will not lock into the BNC body, and you'll get to cut away all your work and try again. Here are some pictures to show a correct crimp.

Figure 3.8.28   Crimp center pin.

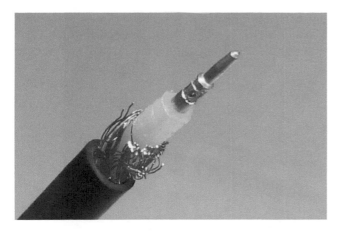

Figure 3.8.29   Good crimp – 1.

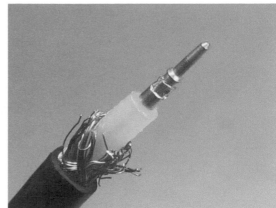

Figure 3.8.30   Good crimp – 2.

Figure 3.8.29 is a front view of a good crimp. Note the position of the crimp on the center pin. Let's see this from another angle. A rear view of a good crimp is shown in Figure 3.8.30 – it's crushed just enough, not too much.

Now is a *very* good time to slide the crimp ring on, and move it far enough down the wire to be out of the way (Figure 3.8.31). Do *not* forget this! If you mount the center pin and find you forgot the crimp ring, you will then have to poke out the center pin – which weakens the pin's lock-in onto the BNC body.

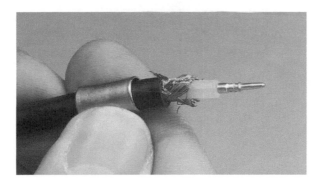

Figure 3.8.31   Mount crimp ring.

Push back the braid toward the outer insulation jacket of the wire (Figure 3.8.32). This will permit the braid to be subsequently brushed forward over the rear of the BNC body, and then mashed down with the crimp ring.

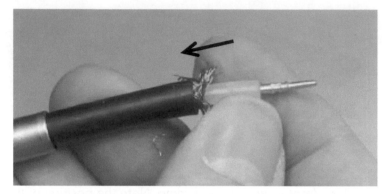

Figure 3.8.32   Push back braid.

Slide the inner insulation into the center of the BNC body (Figure 3.8.33). Align the center pin with the hole in the front of the BNC body, and push the pin through the hole.

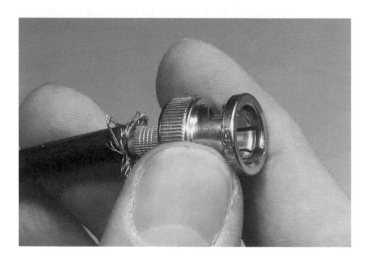

Figure 3.8.33   Mount BNC body.

Gently pull on the center pin with a pair of pliers (Figure 3.8.34). You should hear/feel a sharp 'click' when the pin snaps into place. When that happens, you're done – don't pull any more.

Brush the braid forward toward the front of the connector (Figure 3.8.35).

Figure 3.8.34   Pull center pin.

Figure 3.8.35   Brush braid forward.

Slide the crimp ring toward the front of the connector (Figure 3.8.36). Push it as far forward as possible.

Figure 3.8.37 shows the crimp ring in position. We are now ready for a date with our crimper.

Figure 3.8.36   Mount crimp ring.

Figure 3.8.37   Ring in position.

The crimp ring is wider than the jaws of the crimper (Figure 3.8.38). So the ring crimp has to be done in two parts. I like to crimp the rear first, to lock the ring in position. Figure 3.8.39 shows the details of the first crimp, to show the degree of compression.

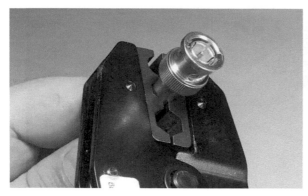

Figure 3.8.38   Rear of crimp ring.

Figure 3.8.39   Rear crimp done.

In Figure 3.8.40 I'm caught right in the middle of the second ring crimp. Note that the crimper's jaws are not yet fully closed. Squeeze, squeeze!

Figure 3.8.40   Front of crimp ring.

Notice in Figure 3.8.41 that I was good and lined up the front and rear crimp. This BNC is done! But before taking it out for battle, there's one more step to do.

Figure 3.8.41   Ring crimp done.

In Figure 3.8.42 the BNC connector is shown being sprayed with contact enhancer. Spray the tip and inside the lock-ring. Wipe off any excess. I'm using Caig ProGold G5 and *you* should too! It improves electrical contact on all metals used in electrical connections, and is especially good for low-power connections.

Figure 3.8.42   ProGold spray.

The signal level on a BNC connector can range from medium strong to very weak, so any help you can give those hard-working electrons will be greatly appreciated!

Note: Caig Labs has recently changed the name from ProGold to DeoxIT Gold. It's the same stuff. They also have some new products that are interesting. More info about ProGold is on the Caig Labs website (www.caig.com).

Now it's time to whip out that VOM you bought, after I told you to do so in Section 1. If you really read the whole book, you'll recall a mini-course in how to use it (the VOM) that was part of Section 2. You'll even remember what VOM, or DVOM, stands for.

Test your work and check it for shorts. If it all checks out, enjoy your new connections!

F male connectors

The F connector is commonly used for video and RF work. It's an unbalanced connector, with only hot (tip) and ground (shield). Unlike most connectors, this one uses the center conductor of the coaxial cable as a pin contact. This creates long-term problems, as the center conductor is bare copper and corrodes easily. So if your old F connector goes intermittent, scrape the oxidation off the center conductor and try it again.

Don't ask me what the 'F' stands for, unless it's the 'f' expletive uttered by technicians when the center conductor corrodes and the contact degrades. Perhaps it stands for 'fairly', as the F connector is fairly cheap, fairly simple to wire, and fairly reliable.

Since the F connector gets a passing, if not an outstanding, grade, it's a common connector on TVs, some video games, closed-circuit camera installations, and other places where video or RF wiring is needed.

The F connector is similar in use to the BNC connector covered in Section 3.8, and the same type of crimper (with different jaws) is used for both.

The differences in usage between BNC and F connectors are discussed at these URLs:

* http://www.l-com.com/multimedia/tips/tip_75ohm.pdf
* http://www.l-com.com/multimedia/tips/tip_50ohm.pdf

Typically, the 50 ohm connectors are BNC, but the 75 ohm ones are often a mixed bag of BNC, F and RCA connectors.

Let's look at a few pictures to show you what I'm talking about.

Figure 3.9.1   Connector rear – 1.

Figure 3.9.2   Connector front – 1.

Figure 3.9.1 shows the rear of the connector, where the wire is inserted. Note that the construction is basically two tubes or cylinders. They are connected at the front of the connector and left open at the rear, to permit a pushdown onto the wire. At the front end of the connector, there is a captive, threaded ring, which is used to secure the male connector to the female. The ring spins freely, so it can be tightened by hand. This is shown more clearly in Figures. 3.9.2 and 3.9.3.

In Figure 3.9.2 we can see the threading in the free-spinning lock-ring, as well as the hexagonal ridges on the ring to make using it easier. The circular ridges on the body (barrel) of the connector add strength to it when it's crimped (gently crushed) down onto the wire.

Figure 3.9.3 is another view of the connector, from a slightly different angle and with different lighting. I hope that from these two views, and my deliberately detailed description, you will now have a clear understanding of the anatomy of a male F connector shell. The clever part of its design is that there are only two parts – the shell and the wire itself. The less clever part is, as mentioned, the use of bare copper for a contact pin.

Figure 3.9.3   Connector front – 2.

Figure 3.9.4 provides one last look at the rear of the connector, carefully lit, so you can see that the space between the inner and outer tubes (cylinders) extends all the way down to the front of the connector.

Figure 3.9.4   Connector rear – 2.

Figure 3.9.5 shows how the shell is used. It has been forced down, over some carefully prepped wire, so that the center conductor, and its surrounding insulation, are properly positioned. Notice that the body of the connector is still round. This means the connector has not yet been crimped, and it's easy to pull the shell out of correct alignment with the wire if you mishandle it. Be gentle.

Figure 3.9.5   Wire in place.

Figure 3.9.6 Two coaxial wires.

Two typical examples of coaxial wire are shown in Figure 3.9.6. Note the single inner conductor, with a thick insulator around it. Also note that the wires are different diameters, so they need F-connector shells with different diameter 'inner' tubes to fit them properly. We'll talk about that more in a minute, but for now I want you to observe that the wire on the left has only a braided shield conductor, but the wire on the right has an additional foil shield. This provides more complete shielding, at a bit more cost, and creates a stiffer wire.

We'll be working with the double-shielded version, as it's the one with a few extra steps. For wire that has only a braided shield, just omit the steps needed to deal with the foil shield.

F connectors and wire come in various sizes and standards. There's the older RG-59U, the thinner (newer) RG-6U, and a bewildering array of other types and sizes. The important thing to check is that the diameter of the inner insulator and the diameter of the inner tube of the F connector are matched. You shouldn't take some clerk's statement that 'They'll fit perfectly' as truthful – the clerk might not know. Peel back a bit of the shield and outer jacket, and physically check the fit of the inner insulator to the inner barrel of the F-connector shell. See Figure 3.9.5 for an example of a correct fit.

There are also a number of different designs of F-connector shells on the market. I can't show all of them to you. The example I use is for the most common one. Use it as a basis for your own work if you are dealing with a type of F connector that is slightly different in its construction.

There are also 'twist-on' F connectors on the market – don't use them. The F connector is mediocre enough without using a type that was developed to avoid the cost of buying or renting a crimper. The 'twist-on' type is both physically and electrically weaker.

For the type of F shell I'm using, I need to start by cutting away the outer insulating jacket (in this case, the orange covering), for a length of {3/8} inch (Figure 3.9.7). Measure the distance, and put your thumb against the outer jacket to guide a razor blade to the exact spot you need, as shown in Figure 3.9.8.

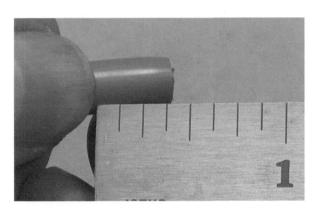

Figure 3.9.7   First cut – 1.

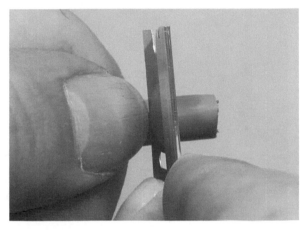

Figure 3.9.8   First cut – 2.

Place the razor blade in position. Keeping the blade at a right angle (perpendicular) to the wire, spin the blade gently around the wire to slice through the outer jacket. Do this lightly, so you do not cut any of the braided strands of the shield below the outer jacket.

If you've been good, and cut gently, you should now have a thin slice through the outer jacket, that goes completely around the wire (Figure 3.9.9). If the outer insulation is not stuck to the strands of the shield, you can just pull the {3/8} inch piece of outer jacket off. If the jacket adheres to the shield strands, go on to the next step.

Figure 3.9.9   First cut – 3.

Only if needed, make a second cut in the outer jacket – from the point of the first cut to the end of the wire (Figure 3.9.10).

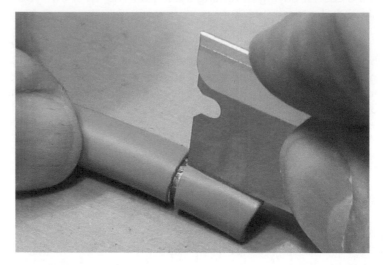

Figure 3.9.10   Second cut.

Peel away the {3/8} inch bit of outer jacket and discard it (Figure 3.9.11). The next step is to cut back the outer shield.

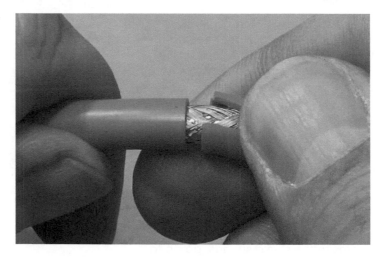

Figure 3.9.11   Remove outer jacket.

Cut away the braided shield (Figure 3.9.12). Cut as close to the outer jacket as possible. Don't worry if you have to make a couple of 'passes' around the wire to get a tight, clean cutaway.

Figure 3.9.12   Remove shield – 1.

If you have the mixed blessing of an inner foil shield, remove it as well (Figure 3.9.13). The same concept applies – keep the cutaway as tight as possible, and don't be afraid to go back and clean it up.

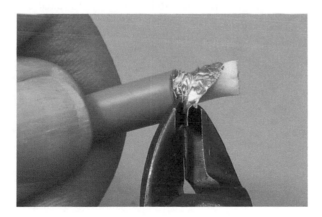

Figure 3.9.13   Remove shield – 2.

Figure 3.9.14 shows the wire after the shield is cut away. I still have to clean up the fragments of shield that are left, but I'll wait to do that until after the next cut. Why? Well…

The next step is to cut the inner insulator back (Figure 3.9.15), which will also leave more room to remove those pesky nubbins left over from removing the shield. Slice gently and try not to score the inner conductor. After cutting it loose, remove the {3/8} inch piece of inner insulator.

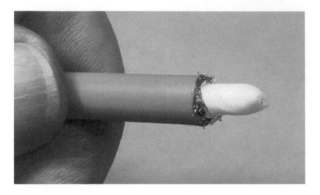

Figure 3.9.14   Remove shield – 3.

Figure 3.9.15   Remove inner insulator.

Wiring is iterative – you get to keep (re)doing it until you get it right. That's OK – it's just part of the process. In Figure 3.9.16 I'm cleaning up the loose strands and spoodge left over from my rough cut-off of the shield. I'll have another go at the inner insulator as well, to make the razor cut across it nice and even.

Figure 3.9.16   Trim shield more – 1.

Almost done trimming! Any one of those loose strands you can see in Figure 3.9.17 is a potential short. So I have one more pass at trimming this connector's 'hair' before I can go on to the next step.

Figure 3.9.17   Trim shield more – 2.

Oops, almost slipped one by you here. Look carefully at the shield in Figure 3.9.18 and you'll see it has finally been trimmed into submission. All this sounds very tedious – trim, retrim, re-retrim. But once you get the hang of it, each operation only takes a few seconds. I now have {3/8} inch of bare, happy, inner conductor, my shield strands are snipped snugly and the inner insulator is now neatly nipped flush across. All is joyful, and we're ready to proceed to the next trim action.

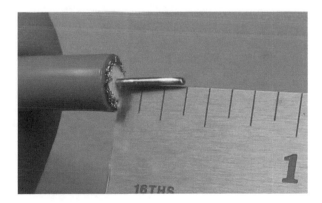

Figure 3.9.18   Trim 1 finished.

Measure back {1/4} inch from the edge of the first trim (Figure 3.9.19). Here, we will *repeat* the exact same steps I showed you for the first trim of the outer jacket (to be exact, Figures 3.9.8–3.9.11). That is, lightly cut the outer jacket and remove the {1/4} inch piece. Then cut away the shield, up to the cut-off of the outer jacket. Figure 3.9.20 shows the jacket cutaway.

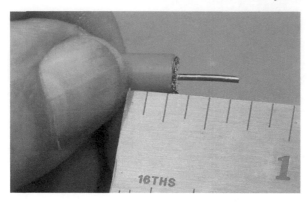

Figure 3.9.19   Trim 2 measure.

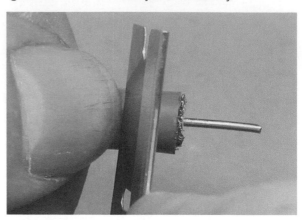

Figure 3.9.20   Trim 2 outer jacket.

Just as we did before, slice the outer jacket delicately and remove the {1/4} inch piece. If you're lucky, you can just pull it off without a second cut. If not, cut it as in Figure 3.9.10.

Cut away the shield, both braided and foil (Figure 3.9.21). This will leave a {1/4} inch 'shoulder' of the inner insulator. It is this shoulder that goes down the center tube of the F-connector shell, supporting and insulating the center conductor. Form it carefully.

Figure 3.9.21   Trim 2 cut shield.

Mea culpa. I should have a picture of the finished trim. But I messed up, and Ken (my photographer) is not around. So the little drawing in Figure 3.9.22 is to try and make amends. This is a simple 2-D drawing of our admittedly 3-D wire, where the big orange part is the outer jacket, the white part is our {1/4} inch shoulder of inner insulation, and the thin copper-colored part at the right is, you guessed it, the inner (copper) conductor. And look, not a stray strand of shield in sight. Nice trim job!

Figure 3.9.22   Cross-section.

The wire is now ready for the F-connector shell to be attached and crimped. But to crimp (crush in a controlled manner) we need a crimping tool, right? We sure do! So if I've got the quote right, 'Let me introduce you to my little friend…' (Figure 3.9.23).

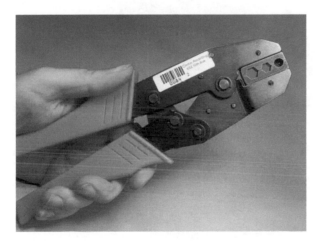

Figure 3.9.23   The crimper.

This type of crimper can be used for a variety of plugs, including BNC, F and Ethernet connectors. The Jaw-sets mounted in Figure 3.9.23 are for BNC connectors.

This crimper is versatile, durable, and has a powerful set of jaws! It's got a crushingly powerful personality, and delights in crunching various plugs into obedient conformity!

Speaking of those jaws, a crimper will only release after full compression, so *avoid getting any part of your anatomy caught in the jaws*. You will either have to let that part get squished, bloody, and mangled, or take the crimper apart to release yourself! The same goes for anything else you put in the crimper, so be careful! Powerful tools are always dangerous. Use them with respect and good judgment.

To be utterly clear, I've drawn two arrows pointing to the thumbscrews on the opposite side of the crimper's jaws in Figure 3.9.24. These thumbscrews are removed and replaced to change jaw-sets. In fact, I did change the jaw-sets – the ones in this picture are for BNC work. Just trying to keep you alert!

Figure 3.9.24   Jaw-set thumbscrews.

Figure 3.9.25 shows a cornucopia of jaw-sets! The BNC set is mounted, the Ethernet set is on the lower left, and the F-connector set on the lower right.

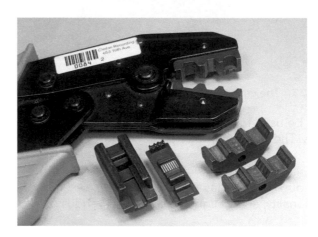

Figure 3.9.25   Multiple jaw-sets.

In Figure 3.9.26 I'm removing the BNC jaws, so I can load the F jaws (jaws of F?).

OK troops, the F jaws are locked and loaded (Figure 3.9.27). Time to rock and roll!

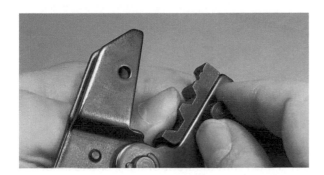

Figure 3.9.26 Removing jaw-set.

Figure 3.9.27 F jaws installed.

Take the F shell, and push it firmly down onto the prepped wire (Figure 3.9.28). The goal is to wedge the inner tube of the shell *between* the inner insulator and the shield. This, and the next operation, may require a little elbow grease, but not too much. Let the shoulder of the inner insulator slide down inside the inner tube of the F shell toward the front of the connector. Then keep going.

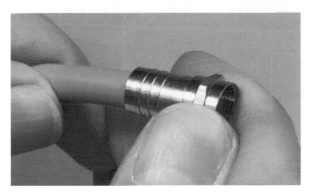

Figure 3.9.28 Push shell on wire.

Push the wire up into the connector – or push the connector down. Or a bit of both. A little wriggling and twisting often helps (Figure 3.9.29). You will know that the wire is correctly positioned when the inner insulator's shoulder is flush with the front of the inner tube in the F shell. This is shown in Figure 3.9.30.

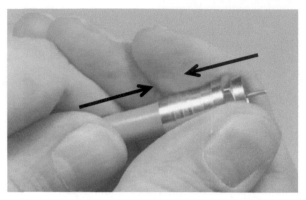

Figure 3.9.29   Seat shell on wire.   **Figure 3.9.30**   Correct positioning.

Remember this one? But now it makes (I hope) a lot more sense. As described, the inner insulation is flush with the front of the F shell's inner tube. This connector is ready for his date with the crimper.

In Figure 3.9.31 the F shell is in the crushing embrace of our crimper. Ow, that hurts! There are several caveats here. When in doubt, start with the widest jaw setting and work your way down! You can always crimp more, but clearing the F shell out of the crimper, when you've caught it in a too-small jaw space, will raise your blood pressure exponentially.

Figure 3.9.31   First crimp.

The same applies to making sure that you are crimping *on* the crimpable area of the F shell. There is sometimes a small section just after the lock-ring that is too thick to crimp. It will also give you severe agitation if you bite down on it (or the lock-ring itself) and find it too tough to crimp.

Note that I'm crimping first as close to the front of the F shell as I can. Then I'll move the F shell to the right, and crimp again to finish it. This is mandatory, as the F shell is wider than the crimper's jaws.

In Figure 3.9.32 I've moved the F shell to the right, so that the uncrimped part at the rear is now inside the crimper's jaws. A second compression stroke and the connector is done!

Figure 3.9.32   Second crimp.

Our finished male F connector is shown in Figure 3.9.33. Isn't he handsome? Note how the marks from the two crimps line up, and the inner conductor extends past the lock-ring – its length is not too critical. Just one last operation to do.

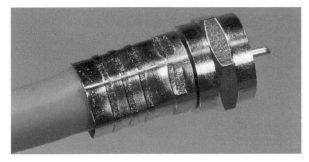

Figure 3.9.33   Finished connector.

In Figure 3.9.34 the F connector is being sprayed with contact enhancer. Spray the tip and inside the lock-ring. Wipe off any excess. I'm using Caig ProGold G5 and *you* should too! It improves electrical contact on all metals used in electrical connections, and is especially good for low-power connections.

Figure 3.9.34   ProGold spray.

The signal level on an F connector can range from medium strong to very weak, so any help you can give those hard-working electrons will be greatly appreciated!

Note: Caig Labs has recently changed the name from ProGold to DeoxIT Gold. It's the same stuff. They also have some new products that are interesting. More info about ProGold is on the Caig Labs website (www.caig.com).

Now it's time to whip out that VOM you bought, after I told you to do so in Section 1. If you really read the whole book, you'll recall a mini-course in how to use it (the VOM) that was part of Section 2. You'll even remember what VOM, or DVOM, stands for.

Test your work and check it for shorts. If it all checks out, enjoy your new connections!

Index